Robert A. Wesolowski • Anthony P. Wesolowski
Roumiana S. Petrova

The World of Materials

Springer

Robert A. Wesolowski
St. Joseph High School
Plum, PA, USA

Anthony P. Wesolowski
Fu Jen Catholic University
New Taipei, Taiwan

Roumiana S. Petrova
New Jersey Institute of Technology
Newark, NJ, USA

Additional material to this book can be downloaded from https://www.springer.com/us/book/9783030178468

ISBN 978-3-030-17846-8 ISBN 978-3-030-17847-5 (eBook)
https://doi.org/10.1007/978-3-030-17847-5

This Springer imprint is published by the registered company Springer Nature Switzerland AG
The registered company address is: Gewerbestrasse 11, 6330 Cham, Switzerland

Introduction

Being involved in education for the past 42 years, I have seen many trends come and go. As to whether or not the education system has become better as a result of these trends is still dubious. In the late 1980s through the early 1990s, a concept of *hands-on* learning transformed the education system with a concept labeled *Principles of Technology* touted by the Center for Occupational Research and Development (CORD). In 1984, while teaching physics at Taylor Allderdice High School in Pittsburgh, Pennsylvania, I had an opportunity to integrate a Scholars Physics Course with an Electronics Course instructed by a colleague, Mr. Edward Karsin. The major purpose of the endeavor is to allow students in the academics share the experiences of the vocational environment. Students were instructing students on the use of electronic equipment. The teachers were facilitating and not lecturing. This peer interaction helped solidify concepts being taught while raising the self-esteem of the student instructor. Prior to this emerging concept of "new hands-on" approach in the education arena, Mr. Karsin and I had already been engaged in such an activity for nigh onto 10 years. We acknowledged and valued its merit and significance, presenting our findings at numerous convocations such as: SREB (Southern Regional Educational Board) Conventions in Atlanta, Georgia, and in Baltimore, Maryland, and NSTA (National Science Teachers Association) Convention, before the Department of Education, in Harrisburg, Pennsylvania.

In 1998, I became involved with the Department of Materials Science and Engineering, Carnegie Mellon University. Interacting and collaborating with Dr. Gregory Rohrer, present Department Chairman, I coordinated a program which created an internship involving high school teachers engaged in creating activities involving materials science, which would be used within their disciplines. With the support of the NSF, the internship existed for approximately 14 years. This activity allowed concepts of materials science to be introduced and integrated into a high school curriculum. My initial activity with Dr. Rohrer, which involved research with the torsion pendulum in a static state, allowed me to transfer an applied theory to a hands-on experience. Developing a static torsion test to prove the relationship between force, radius, and the angle of rotation of a twisting material, with the help of various professors at Carnegie Mellon University, introduced ideas of how this device could be used to explain a three-dimensional concept with angular motion. In September, students amazed me with their curiosity of my summer work. As a result of the student's curiosity into this arena, this summer project expanded into a student project that was presented at PJAS (Pennsylvania Junior Academy of Science). "Hands-on experiences" do work. Furthermore, the integration of science and mathematics with this hands-on experience stimulates a student's curiosity and encourages learning. During this time, I was fortunate to see other teachers developed their idea integrating materials science with their curriculum.

In 2008, I was approached by Dr. Anthony Rollett, Professor at Carnegie Mellon University, to assist in bringing an ASM International Materials Science Teachers Camp to Pittsburgh, Pennsylvania. This camp was to expose teachers to what materials science could afford their students in a *hands-on* environment. The first camp hosted 35 teachers. These teachers were introduced to concepts taught in other scientific disciplines, demonstrating their applicability to the world of materials science. Their experiences were positive, and it was found that 42% surveyed used at least one or more of the learned activities in their science curricula. As the years progressed, I became more involved in the ASM Teachers Camps and eventually became a master teacher traveling around the country instructing teachers in the method of incorporating materials science into their syllabi.

It has been brought to my attention that the Pittsburgh Public Schools were interested in developing a materials science course for their high school students. Research suggested that there was a need for a unified curriculum. This I proposed to Carnegie Mellon University, ASM International, and the Pittsburgh Public Schools. Through hard work, a skeletal course was presented to the school district and a materials science course was presented. This course has since been modified and expanded into a semester course and a year-long course. However, a challenging textbook was needed. I took it upon myself to begin writing a textbook, supplying needed explanations which would assist and ensure instructor success.

The next adventure will be to create a website available to materials science teachers providing a forum wherein instructors, interacting with one another, would enable participants to gain understanding and insight. Too often, teachers are left to their devices, isolated and with little direction other than the textbook and a "curriculum." With website intervention, the instructor need no longer feel isolated.

I hope the intent of my writing is fruitful, and the book and the curriculum are successful in helping enrich our students.

St. Joseph High School, Plum, PA, USA Bob Wesolowski

Contents

Chapter 1
Introduction to Materials Science

As an individual works their way through the educational system, topics experienced, range from topics that fall in the realm of liberal arts to the sciences. Each topic is a stand-alone, unique to itself. At times there is a dependency on previously taught material as experienced in a mathematics course, but moreover, individuals seldom recognize the integration of one subject into another.

Oobleck: White Powder Lab

Living in our world, materials experienced usually fall into categories of solid, liquids, and gases. At least this is what was taught to us years ago. Later a presentation of a material called plasma was introduced. It was easy to comprehend a solid because it had structure, rigidity, and density. When liquids were introduced, the idea of viciousness was also introduced describing the flowing of this liquid material and comparing it to other liquids. Gases presented a different situation because they were light and invisible under normal situation. This highly active material surrounds us and can be produced by applying heat to a light material. The introduction to plasma came years later because it was a by-product of heated gas. A plasma is created by heating a gas or placing a gas in the environs of a strong electromagnetic field having been applied via laser microwave generators. The electrons of the gas change as a result of this interaction creating either a positive or negative charged particle called an ion. At times materials behave in a manner unique to itself and do not fall into a simple description of a solid, a liquid, or a gas. Consider a material such as a polymer. A polymer is a term composed of the part **poly** which means many and **mer** which means a unit. Therefore a polymer is a many unit material, existing in a chain so long that this chain interferes in the flow or viscosity or the material. Consider the materials of honey and corn syrup which can result in a very slow flowing material as compared to water. In the lab being performed, an understanding of the states of matter is important in order to investigate the properties of the mixtures created. Other terms that may be helpful in understanding are solutions, colloids, and suspension. A solution is a liquid mixture combining a solute a

R. A. Wesolowski et al., *The World of Materials*,
https://doi.org/10.1007/978-3-030-17847-5_1

Fig. 1.1 Virco 9000 series student desk w/ bookrack. The illustration leads to the discussion of the complex make-up of the composition of the student desk

minor component to a solution a major component, such as Kool Aid and water. A colloid is a substance made of a material being suspended in another material. A solution is a combining of substances; a colloid mixes materials but do not combine the material, such as an introduction of a gas to a cream. A suspension is a heterogeneous material or larger mass that are floating around in another material. These materials do not dissolve. Keep these concepts in mind as we proceed with the Oobleck lab.

As an introduction to a course defined as materials science, an understanding of the word "material" is essential. What makes the "material"? Does altering the "stuff" that makes the material change the original material? Knowing what was the change and how did the stuff affect this change is critical. So, what is this "stuff" that makes a material and how can it be modified to alter its properties? A few questions that can help in the understanding of materials are: Where did this material originate? How is this substance made? Why do we use the particular stuff in the making of this "material?" Is the "stuff" identifiable? Make a list of materials and the "stuff" used to make the material. What is the composition of the material?

Consider a classroom desk (Fig. 1.1):

Consider the amount and types of metals and plastics used in this desk. Chrome-plated steel tubes provide the basic structure followed by plastic or composite seats or desktops. Steel fasteners such as rivets, screws, nuts, and bolts are used to hold the item together. Why were these materials selected? Let's consider durability, strength, and weight. Is there anything else? How easy was it to manufacture these products? Did the material used affect the appearance of the product? And most significant, consider the cost? The purpose in manufacturing a product is making a profit. The absence of a reasonable profit will result in the discontinuance of production of that product.

In essence, what is materials science? Materials science is the design and development of new materials, explicitly ceramics, composites, metals, polymers, and electronic materials. Elements of both the physical and the chemical composition of these materials need to be investigated. We cannot study materials in a vacuum. Materials science is an interdisciplinary study involving physics, biology, chemistry, mathematics, applied arts, and technology. A "material scientist" grows new materials, makes new use of old goods, and determines how to process resources into useful components. "Materials science" is an integral component of manufacturing. The industry focuses on the proper use of the given resources. The automobile industry was compelled to increase mileage in the car; vehicles were redesigned with more plastics, more aerodynamic styling, and modifications to the engine. These innovations led to a creation of the electric car. The electric car is not a new innovation but an old idea altered as a result of newer technology. From where did the new design of the battery come? Consider other elements we take for granted in this little world of ours.

The cellular phones have had an enormous impact on our lives, our culture, and our society as a whole. What about the impact cellular phones have had in development of miniature computers? Is our clothing not immune to development? Have you heard of cold wear or hot wear heating devices that keep our bodies warm when we hunt, or ski, or merely walk in winter? There have been radical changes in sporting equipment. Can you name a few changes in the material used in the sports you participate? Are the materials we employ newfound or a combination of older materials? How were these elements selected? Look at the following website: http://www.strangematterexhibit.com/

Considering the aforementioned facts may aid in our understanding, use, and investigation of materials that enhance a product. But from where do the materials we use come? The materials are found in the earth. Consider:

1. Metals extracted from specific ores.
2. Polymers derived from petroleum.
3. Fiber glass created from specific ores.
4. Recycling of materials.
5. Composites a combining of a multitude of materials.

How do the scientists utilize this "stuff?" Scientists improve discoveries by revealing or innovating new means by which these materials can be used. Remember that scientists and the engineers must depend on materials; the scientist discovers new worlds, and the engineer designs new worlds. Manufacturing depends upon improvements and development of these new materials.

What is MSE? MSE is a branch of engineering. It stands for materials science and engineering. It is the only engineering area that also considers itself a science. The field encompasses a multitude of areas:

- Metals.
- Ceramics.
- Polymers (plastics).
- Semiconductors.
- Combined materials known as "composites".
- Biomaterials.

Materials science relies heavily on chemistry and physics, but there is also a reliance on the fields of biology, mathematics, the arts, and technology. Two new fields in materials science have evolved: biomaterials and nanomaterials. Can you think of how materials are applied in these areas? Consider the following devices used in the medical community: the titanium hip, the ceramic socket, the pacemaker, the ceramic valves, the stints, and the plastic skin. Are there any other advances in medicine that have new materials being applied?

Nanomaterial is a new world. We are talking about the size of ~10^{-9}. There is exploration being done in cancer research where stints, the size of 10^{-9}, are having metals placed in them size 10^{-12} and have these stints interact with vibrational motion. This vibration causes molecular motion which then produces heat. When the heated stint of metal shavings is placed in a cancerous tumor, the tumor is killed, but the heated stint does not affect the healthy tissue. Are there other areas when the nano-world has been introduced to materials science?

"Materials science" investigates the relationship between the following components: structure, properties, and processing. Structure relates to the scale of the material. Investigations can be in the atomic size. Here the investigation is centered on the assembly of the atoms and how the atoms are bonded (i.e., joined). The investigation can expand to clusters within the realm of nanometers to micrometers or the investigation found in the macroscopic world ranging from the size of a millimeter to mega meters (10^{+6}).

Investigating properties of materials enables the scientist to describe the materials' physical attributes: how they look, feel, and act. One noticeable property of any material is that it occupies space and possesses mass. The atoms found in materials move around differently from one material to another. Those particles (i.e., atoms) that make the solid do not move around and are usually aligned in a structured pattern. When those particles are heated, they will vibrate. Their vibrations increase as more heat is applied or their movement decelerates as heat is removed. The particles of liquid have movement, and this movement is initially in a rotational manner. The warm liquid rises to the top pushing the colder liquid to the bottom. The movement of a liquid becomes more chaotic as more heat is applied. The gaseous state of matter possesses the largest amount of kinetic energy. The kinetic energy results in the movement being aggressive and possessing the greatest velocity.

In the realm of materials science, the term "possessing" refers to the ability of observing qualities, measuring quantities, sorting and/or classifying, inferring, predicting, experimenting, and communicating. The concept of "processing" is the initial step to problem solving, questioning, investigating, hypothesizing, and evaluating the hypothesis through experimentation; analyzing data and drawing conclusions are modes of a scientific methodology process. Finally, communication of the results is needed.

Let's explore products indigenous to materials science studies. Have you heard of a product called graphene? Graphene is a single layer of carbon that is bonded together in a multiple pattern of hexagons. In 2002, at the University of Manchester, a researcher Andre Geim challenged a PhD. student to polish a piece of graphite to its smallest possible size. After successfully polishing the graphite to a very small size, Geim laughed, tossing the material into the trash where it landed on a piece of tape. The investigation indicated that the tape lifted off a layer of the graphite. This process continued until the result was a two-dimensional crystal, actually, the first two-dimensional crystal. Considering the state of the polished graphite, it was deemed that graphite would be an unacceptable name in describing this new material. Thus "graphene" was adopted as its new name. Having created this material was the first step. Upon analyzing the material, it was found that the electron movement in the material was faster than the movement in silicon. A replacement for silicon was the hope of the computer industry; graphene may be the answer. It is stated that the electron movement of graphene was faster than silicon, but by what amount? It was found that the speed of graphene was 250 times greater than silicon. In addition, the material was stronger. This graphene was 150 times stronger than steel, but it was pliable. The application of graphene is unlimited. The battery may be redefined. Remember the term graphene. It will be coming to a store near you! (Colapinto, December 14, 2014).

[Since the initial publication of the *World of Materials*, the discussion of graphene went from a new discovery, which had properties that were indeed unique, to a technology that is expanding in all direction. From an article credited to the University of Manchester, the "home of graphene," we know.

> Graphene: the world's first 2D material. Since graphene's isolation in 2004 it has captured the attention of scientists, researchers and industry worldwide. This material is
>
> - ultra-light yet immensely tough.
> - 200 times stronger than steel, but it is incredibly flexible.
> - the thinnest material possible as well as being transparent.
> - a superb conductor and can act as a perfect barrier – not even helium can pass through it.

It has been found that graphene is being used in desalination. "Clean drinking water for millions. Graphene membranes could see huge progress in water purification technology in developing countries and provide more efficient desalination plants.

Electronics and energy storage could also be revolutionized by graphene. It is

- flexible,
- durable,
- semi-transparent mobile phones,
- Wearable technology,
  - clothing that communicates.
- Electric sports cars.
- Lightweight planes.

These are the future technologies which are becoming realistic in our present." (http://www.graphene.manchester.ac.uk/explore/what-can-graphene-do/).

As discussed in an article in the International Weekly Journal of Science, Volume 503 Issue 7476, November 20, 2013, "Graphene could… be useful in medical and security scanning that uses terahertz-frequency radiation." "Others think that graphene's most obvious optical property — its transparency — may yield its first major application in the electronics industry. Samsung and other Asian companies are developing transparent graphene electrodes to serve as smartphone touchscreens."]

Let's consider LiquiGlide. Five Massachusetts Institute of Technology (MIT) students and their Professor Kripa Varanasi came up with a material that would reduce friction of liquids. Ironically, resistance is not reduced in solids. LiquiGlide adds a microscopic slippery coating on almost any material, can be affixed to materials reducing friction through the introduction of LiquiGlide. LiquiGlide has potential of reducing air resistance in the aeronautic industry (Subbaraman, October 13, 2013).

Rice University had researched and discovered a new battery prototype, a spray on battery. Neelam Singh, Rice graduate student and lead author of the team's report, stated that technology will only improve when coupled with modern methods. "Spray painting is already an industrial process, so it will be very easy to incorporate this into industry," she said. "We really do consider this a paradigm changer." Like any other battery, the anode, the cathode, and the electrolyte are present in application. You can spray it on a surface and you have an instant battery. Can you envision future application with this technology? (De Paula 2012, June 30).

In the biological world, we have seen the process of regeneration. This ability exists and manifests itself also in the polymer world. Self-healing polymers that can spontaneously achieve quantitative healing in the absence of a catalyst are happening NOW! An industrially familiar, permanently cross-linked poly (urea–urethane) elastomeric network has mended itself after being cut in two by a razor blade. "David Mecerreyes, a polymer chemistry specialist at the University of the Basque Country in Spain, sees opportunities to use this elastomer to improve the security and duration of many plastic parts, for example in cars, houses, electrical components and biomaterials" (Polly Wilson, September 13, 2013).

Starlite is a material paste which can be applied to any surface. It claims that this material is capable of sustaining temperatures well within the thousands of degree range. This material with its alleged claims was created by an amateur chemist Maurice Ward during the 1970s–1980s. Starlite received a great deal of coverage in 1993. Despite interest by NASA, Mr. Ward never revealed the composition of the material indicating that family members were aware of the fabrication process. Since his death, Starlite has never been fabricated by any family member. Is this fact or fiction? If it is a fact, consider the applications this material can have on our world (Kennedy 2013, November 25).

When materials are researched, they are categorized into structure and properties. Structure is the organization or arrangement of the parts of the materials and property or characteristic of a substance that is observed, the chemical composition. One such property is defined as a Newtonian fluid. A Newtonian fluid is one that has a constant viscous flow. The force applied is proportional to the change of the velocity vector.

Consider a "white" powder. The powder can be analyzed; feel the powder, and smell it (use the wafting techniques shown in chemistry).

If an amount of water is added to the "white" powder, attempt can be made in mixing the water with the powder. When the stirring is done slowly, the procedure allows the fluid and powder to mix without significant resistance. However when the mixing is attempted in a rapid manner, a substantial amount of resistance is experienced. Further analysis of this water powder mixture can be seen by slowly placing a finger and submerging the finger into the material. When the finger is placed into the mixture rapidly, a substantial amount of resistance is discovered. If the mixture is poured onto a hand, and pressure is applied, the mixture has all the characteristics of a solid. However, once the pressure is released, the mixture becomes a liquid. Interesting!

This material is a non-Newtonian liquid because it manifests characteristics that resemble those of a solid; however, the material reverts to a liquid when pressure is stopped being applied. Can this material have any purpose in our world? How can this or these properties be applied to enhance properties of existing materials or be introduced in new material where pressure applications can change the state from a liquid to a solid. The solidification is referred to as a dilatant material. A non-Newtonian solid material becoming viscous as pressure is applied is called a thixotropic material. Are there any areas where materials could be improved using a dilatant or thixotropic additive? Consider the areas of body armor and paint where these ideas are implemented. Could there be an application of this material to help prevent concussions in specific sports?

What is important is observing the process of analysis in looking at the non-Newtonian fluid and the application of the scientific method in the analysis. "The scientific method is a way to ask and answer scientific questions by making observations and doing experiments" ("steps of scientific"). To mix the material is only one issue. The application of all the senses must be applied while mixing. Is there a difference when the material is being mixed slowly as compared to being mixed rapidly? Is there an odor in the material? When the material is mixed and then rolled in the palm of the hand what was the experience? Was the material produced smooth, intact, or crumbling into pieces? Was the material clinging or sticky when the material was in the palm of the hand? When pressure is applied, does the material maintain its physical appearance or does it change? The steps of the scientific method are:

Ask a question.
Do background research.
Construct a hypothesis.
Test your hypothesis by doing an experiment.
Data and draw a conclusion.
Communicate your results.

The "white" material is a non-Newtonian material, cornstarch. It does not obey the laws of Newtonian Physics. The material has a propensity to become more rigid as pressure or a shearing force is applied. The non-Newtonian material is a polymer being made of a long molecule. As water is applied to the material, cornstarch, the material does not dissolve within the water; rather it is suspended. The material molecules roll up at each end resulting in the molecules rolling about other when the material and water are stirred slowly. When a greater amount of shear force is applied, the molecules uncurl, becoming linear, causing the materials to entangle with each other and become more rigid. When the shear force is reduced, the material rolls back up and moves freely and returns back to the liquid state (Figs. 1.2 and 1.3).

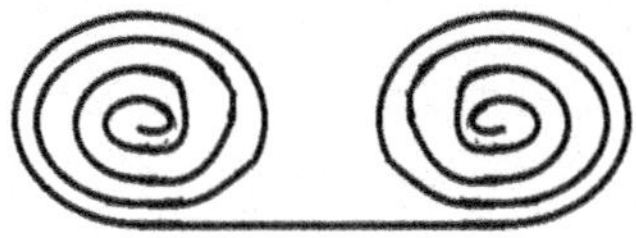

Fig. 1.2 Representation of curled molecules without the influence of a shear force demonstrating the effects when no force is being applied to a non-Newtonian fluid. Everything is Materials, page 21 of solids, ASM Educational Foundation

Fig. 1.3 Representation of uncurled molecules as a result of the influence of a shear force, demonstrating the effects of a force being applied to a non-Newtonian fluid. Everything is Materials, page 21 of solids, ASM Educational Foundation

Welcome to the world of materials science. As the chapters unfold, analysis of stuff will occur. Properties will be investigated, changed, and analyzed to give a better understanding of the stuff, and a potential exists to enhance the "stuff" to better meet the needs of the world of materials.

References

Chemistry World, Polymer regenerates all by itself, Polly Wilson, (2013, September 13).

Colapinto, J. (2014, December 22). Material Question. *The New Yorker*. (Colapinto, 2014)

De Paula, M. (2012, June 12). Soon We Might Spray-Paint Batteries Onto Any Surface. *Forbes*. (De Paula, 2012)

Kennedy, N. (2013, November 25). The Secret Life of Starlite. Retrieved December 27, 2014, from http://www.isciencemag.co.uk/features/the-secret-life-of-starlite/

Subramanyam SB, Rykaczewski K, Varanasi KK (2013) Ice adhesion on lubricant-impregnated textured surfaces. Langmuir. 29(44):13414-13418. https://doi.org/10.1021/la402456c

Srinivas Bengaluru Subramanyam, Konrad Rykaczewski, and Kripa K Varanasi, Langmuir. From a paper: (March 5, 2013).

Chapter 2
Classification of Matter

Understanding the basic concepts usually relies on the ability to break down a topic into its basics. No matter how complicated an object or material is, it can be understood to be composed of so many atoms in a certain arrangement. From this base, the more complicated is evolved.

Classifying Matter

When classifying matter, the material under investigation may be reduced to its atomic structure. The atom is the most basic structure of the material which may be viewed as the element of the material. Copper (Cu), oxygen (O), and tin (Sn) are examples of elements. Combining of elements is a result of bonding covalent, ionic, or metallic. This combines two or more elements to form a compound. This combining can be illustrated by the following compounds: H_2O, NaCl, and CO_2. There are times when there is a combination of elements and compounds. This combination is referred to as a mixture. A basic understanding of mixtures is illustrated when we mix sand and gravel or peas and carrots. Simple Kool-Aid and water are also an example of a mixture. The illustration of peas and carrots is a heterogeneous mixture. It is referred to as heterogeneous because of the ability to recognize and distinguish the components of the mixture. The peas are able to be recognized over the carrots. However the Kool-Aid mixture is an example of a homogeneous mixture. This type of mixture does not allow the differentiation of the components; the final solution of the mixture is a combination of water and the Kool-Aid powder resulting in a liquid Kool-Aid where the components are not unique to each other.

A method to help understand an element from a compound from a mixture is a technique where washers, nuts, and bolts are used. The washers, nuts, and bolts would represent the elements; combining them would be a mixture of elements. If a nut and bolt are screwed together, this would represent a compound. Adding a washer to the compound would simply make a different compound. If these two compounds are combined, the result would be a mixture. Investigate the various combinations and determine whether the respective combination is a mixture or a compound. Are there any elements provided?

R. A. Wesolowski et al., *The World of Materials*,
https://doi.org/10.1007/978-3-030-17847-5_2

In materials science, terms must be clearly defined and understood. Precision in the use of terminology is a must. Therefore, we will begin by defining and differentiating science and technology. Science is defined as "a branch of knowledge or study, dealing with a body of facts or truths systematically arranged and showing the operation of general laws" (Dictionary.com 2015). Technology may be defined as "the heir industrial arts, engineering, applied science and pure science" (Dictionary.com 2015).

Science concerns itself with a search for knowledge and a search for understanding. Technology concerns itself with taking the acquired knowledge of the world and applying that discovered science to the world to improve it. In short, science is discovery and knowledge, and technology is application. When beginning an investigation into materials, this initial concept must be established. Of what does the material (generically speaking) consist? The answer is MATTER. This matter both occupies space and is composed of mass. Mass is a composition of ATOMS, with atoms being the smallest element which contains properties of that element. Chemistry has taught us the various states of matter: solids, liquids, gas, and plasma. Each state possesses uniqueness. Solids possess definite shape and volume. The particles in the atoms have the least mobility. The liquids, with their definite volume, are dependent on a vessel which houses the liquid to provide its shape. The gaseous state is made of particles which possess indefinite volume and shape. The particle movement is the most mobile. Their particles are loosely packed. By combining atoms, through bonding, ionic, or metallic, of the basic element, it makes a compound, and mixing the elements with compounds creates mixtures. A technique used to help visualize the differences in elements, compounds, and mixtures involves the usage of nuts, bolts, and washers. This activity has been posted by a few sources including Flinn Scientific (Fig. 2.1).

Though the nut, bolts, and washers are different, they represent different elements. In a cup, if the nut, the bolt, and the washer are placed, they would represent three different elements. A second cup may have a combination of items screwed together; this combination would represent a compound. The third cup has a combination of various nuts and screws fastened together as well as not fastened together. This combination would represent a mixture.

Reviewing, a mixture is two or more substances physically combined (peas and carrots, Kool-Aid and water, salt and pepper). There is no formula for a mixture, merely a ratio. An element (which is the most basic combination of atoms) can be categorized or grouped into the following: metals, nonmetals, metalloids, and semimetals.

Fig. 2.1 Nuts and bolts: The nuts and bolts are illustration that can be used to describe elements, compounds, and mixtures

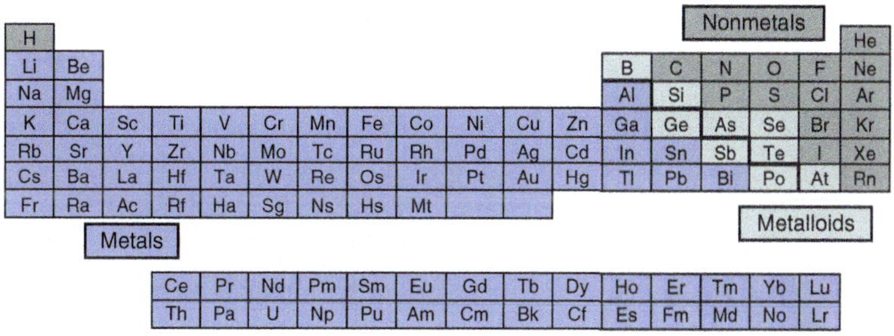

Fig. 2.2 Periodic table indicating the metallic and nonmetallic elements

It is easy to identify a metal from a nonmetal; a metalloid is a nonmetal that when combined with a metal forms an alloy (iron and carbon form steel). A semimetal is a chemical element possessing properties of a metal, but it also may contain properties of a nonmetal. A semimetal can be shiny, or it may be dull. It may be brittle or supple. It may be very hard or soft. It can conduct electricity. The most common semimetal is silicon. Look at the periodic table (Fig 2.2):

Metals are located on the left side of the periodic table; they are numerous, solid, workable, and conduct electricity. Nonmetals are located to the right of the periodic table. The nonmetals can be found in all three states of matter. They are dull, brittle, and tend to be nonconductors. The nonmetals are located on the fringe of the metals and nonmetals on the periodic table. They are the fewest in number in the collection of elements and have intermediate properties. An intermediate property describes a developmental stage transitioning from one type of element to another.

Having identified elements, compounds, and mixtures, and identified metals and nonmetals, has opened the door to an in-depth analysis of a material. An investigation of a substance will focus on the structure of the composition. There are three possible categories of structure for investigation: the atomic, microstructure, and macrostructure. The atomic structure examines the bonding and arrangement of atoms. The microstructure investigates the crystalline or amorphous nature of the structure. Microstructures may be viewed through a microscope. The final structure for investigation is the macrostructure. Macrostructures may be represented by cracks that can be observed with the naked eye.

The microstructure structure refers to the crystalline or amorphous structure. The structure is referring to the atomic structure, in what manner are the atoms arranged. The crystalline structure is an orderly arrangement of particles, long range and predictable (Fig. 2.3).

The amorphous structure is the random arrangement of particles, no repeating pattern, and not predictable (Fig. 2.4).

Consider the table of the classification of materials. Fill in the blanks (Fig. 2.5).

The history of mankind both indicates and confirms early material discoveries and their applications. Which material was used by man the longest? Utilizing the date map while being aware that the map is not drawn to scale, decide which materials were in existence the earliest, and how was the material being used?

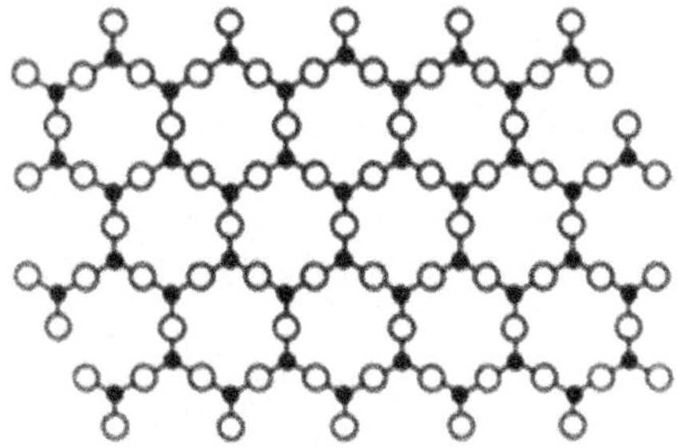

Fig. 2.3 Crystallinie model showing the organized structure that is associated with the term crystalline

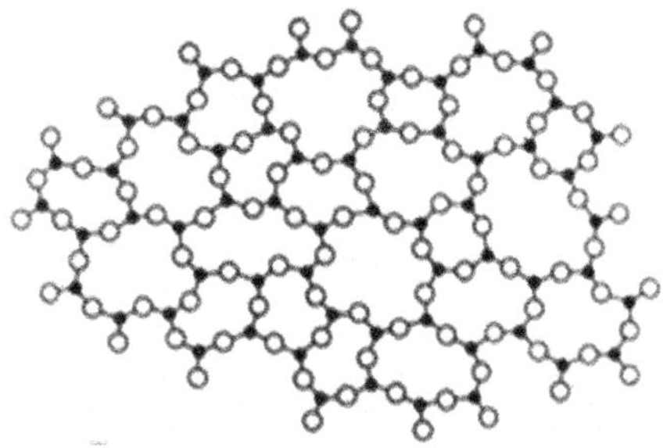

Fig. 2.4 Amorphous model of a crystal. The disorganization or the chaotic structure of the model illustrates an amorphous model

Fig. 2.5 Chart to be filled in to compare the different types of material's composition. Comparison of type of material, bonding, and types of elements that compose the material

Materials classification			
	Metals	Ceramics/glass	Polymers
Type of matter			
Type of elements			
Type of structure			
Type of bonding			

As time marched forward, the quantities of given material, previously used, have changed. Notice the blocks of time where conflict occurred. What materials were used extensively? Refer to history and consider the industrial revolution. What metals were the most used? In the latter part of the 1950s and 1960s, the material of choice was plastics.

This fact regarding plastics was stressed in the 1967 movie *The Graduate* "…the future can be summed up in one word, '*plastics*'" (Fig. 2.6).

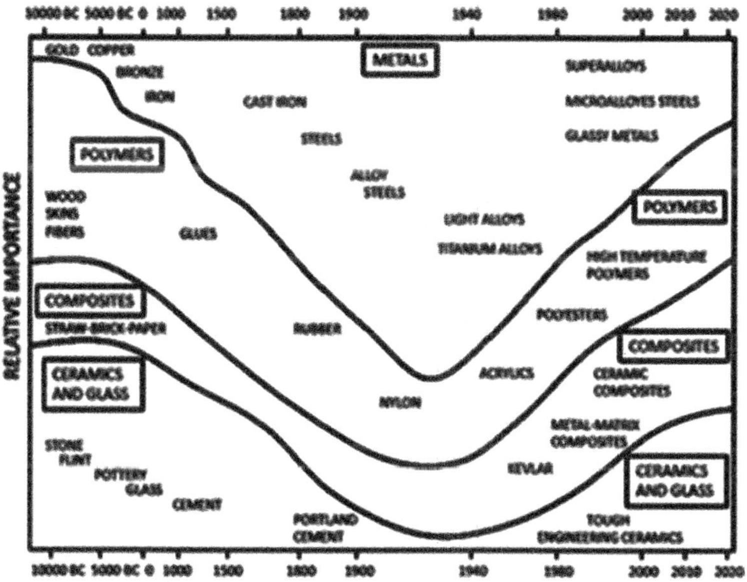

Fig. 2.6 Adapted from Stuff: The Material the World Is Made of, William Morrow, ISBN 0380731533 By: Ivan Amuto. The purpose of the illustration is to show the development of various materials which can lead into discussion of why the particular material was prominent at that time in history. The time is not incremented. Everything is Materials, page 33 of solids, ASM Educational Foundation

An interesting book entitled *The Substance of Civilization* written by Stephen L. Sass on August 11, 2011, has an excellent approach in the analysis of history based on the materials use in the given time periods and how the use of the material helped mold our civilization. Society of today has developed over time and has been affected by a multitude of events, Roman Empire, Catholic Church, Chinese society, World War I and II, and Space exploration to name a few. Many of the events are centered around the availability of product. Metals on the earth have always been available; however the technology to extract the metal from its basic ore structure has not always been known. As advances in society grew so did the applications of product and the creations of new materials.

To understand the composition of material, an awareness of the atomic structure of the material is essential. Consider the structure of the atom. It is to be recognized as the smallest particle consisting of the properties of a particular element. It contains three basic subatomic particles: protons, neutrons, and electrons (Fig. 2.7).

In the small and dense center of an element lies the area called the nucleus. It is small and dense and located in the center of the element, with the proton and the neutron housed in the nucleus. The electrons are rotating around the nucleus. Moving outward from the nucleus are shell structures that locate the electron. The electrons orbit in the shells while spinning upon electron's axis.

Fig. 2.7 Graphical illustration of the composition of an element, illustrating the nucleus, protons, and electrons

Mainly, the composition of an element consists of the mass of the proton and the neutron located in the nucleus. The electrons, which equal the number of protons, orbit the nucleus. In its orbital state, the electron has a minimal mass, compared to that of the element located in the nucleus.

Recall the atomic structure presented in chemistry class. Refer to the atomic mass unit. The atomic mass unit (amu) is the standard applied in measuring the mass on the atomic or molecular scale.

The atomic number, which positions the element on the periodic table, is the quantity of protons contained in the element, as well as the sum of the electrons.

The *atomic mass*, the mass of the element, is the total number of protons and neutrons found in the nucleus. Usually credit the mass to be a whole number. If the element is on the periodic table, the above information can be found as illustrated below (Fig. 2.8).

Remember that an isotope is an element that contains the same atomic number but possesses a different atomic weight or mass. This implies that the number of protons is unchanged. However, the number of neutrons within the nucleus is altered. To determine the number of neutrons, reduce the atomic mass of the isotope by the atomic number, the number of protons. The isotope may be stable, but it may similarly be radioactive.

Complete the chart below for better understanding (Fig. 2.9).

Examining the element beyond the nucleus, the electron is found. The shells housing the electrons contain a definite number of electrons per shell. The sum of electrons is based on $2n^2$ where n represents the shell. However, there is one stipulation. Never place more than eight electrons in the outer shell prior to beginning another shell. Can you determine the electrons needed for a given shell?

An electron configuration determines how electrons are distributed in atoms (Fig. 2.10).

An electron configuration of an atom is the representation of the arrangement of electrons distributed among the orbital shells and subshells.

An electron configuration is based on the quantum model of the atom. The element's configuration is based on the previous element's configuration.

Elements are joined to form compounds using a technique called bonding. Covalent, ionic, and metallic are types of bonding. Furthermore, the Van der Waals forces, the attractive force between molecules, are also used to join elements. The electrons used in bonding are referred to as the valence electrons. These electrons are located in the outer most shell of the electron orbit. The remaining electrons are referred to as core electrons.

Fig. 2.8 Element from periodic table illustration to indicate what information is found on the periodic table for each element

Fig. 2.9 Subatomic particles chart designed to teach "what are the characteristics of the particular particle and its purpose."

Fill out the table based on what you just read

Name	Proton	Neutron	Electron
Charge			
Location			
Mass			
"job"			
Number			

Fig. 2.10 Electron distribution: the representation of the arrangement of electrons distributed among the orbital shells and subshells

Atom	No. electrons	Subshell electron configuration
H	1	$1s^1$
He	2	$1s^2$
Be	4	$1s^2\,2s^2$
B	5	$1s^2\,2s^2\,2p^1$
C	6	$1s^2\,2s^2\,2p^2$
N	7	$1s^2\,2s^2\,2p^3$
O	8	$1s^2\,2s^2\,2p^4$
F	9	$1s^2\,2s^2\,2p^5$
Ne	10	$1s^2\,2s^2\,2p^6$

Investigating the metallic bonding of electrons is experienced when pure metals exist. In a solid, the electrons are discovered encircling the positive ion core like a "sea" of electrons. This electron movement creates an electronic attraction to the solid (Fig. 2.11).

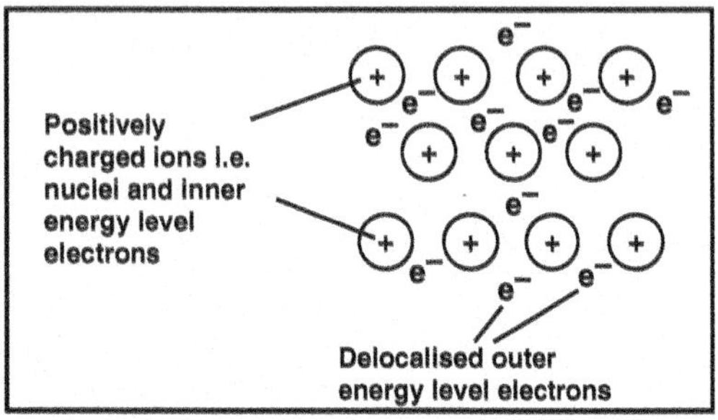

Fig. 2.11 Electron motion and movement in solids; the electrons are discovered encircling the positive ion core like a "sea" of electrons

The valence electrons are delocalized, being shared by all of the atoms of the material. The electrons are free to drift throughout the material. This movement creates a uniqueness found only in metals: the shiny luster and good thermal and electrical conductivity.

Ionic bonding is the process of an electron leaving one element for another. The cation is referred to an atom which yields an electron. The anion is referring to an atom that is a recipient of an electron. Atoms that relinquish an electron, subsequently are deficient of an electron (−), or an atom which accepts an additional electron and therefore possesses an excesses electron (+), are called ions (Fig. 2.12).

Covalent bonding is the sharing of electrons among two or more atoms. The electrons shared are the "s" or the "p" electrons. The created bond is durable but lacks electron movement between atoms.

To reinforce the understanding of bonding, complete the chart below (Table 2.1):

The completion of the chart can be located in Appendix A.

Consider looking at various types of materials, can a decision be made to determine whether the material would be a composite, polymer, metal, or ceramic? What properties would be considered to make the decisions? How do the properties of a metal differ from a ceramic or a polymer differ from a composite? Make a list of properties identifying the characteristics of the materials (metals, ceramics, polymers, or composites) and describe what similar qualities would the materials possess. Types of materials that can be used in the investigation are listed below. Having secured samples of these materials, the students can be given an object from this collection and be asked to identify what the item is, relative to what the student believes constitutes a metal, ceramic, polymer, or a composite. Having made the decision of the type of material selected, a description of the characteristic can be the next challenge. Different materials may be grouped under one category, but do the materials of that same category have the same characteristics? (Tables 2.2 and 2.3)

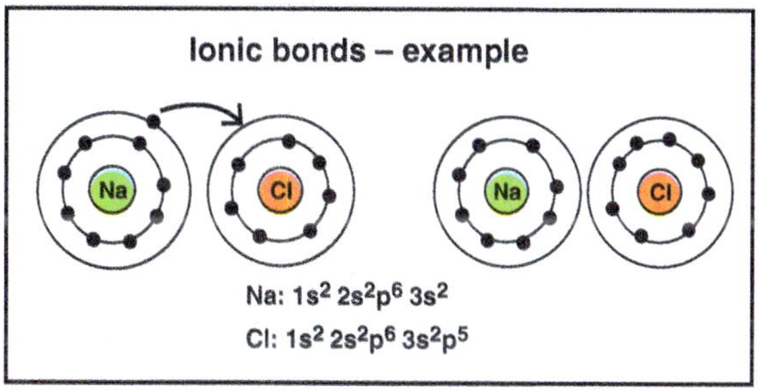

Fig. 2.12 Illustration of a transfer of an electron in making an ionic bond

Table 2.1 Types of atomic bonding chart used to demonstrate the type of ionic material, a giver or a receiver of electrons, strength of the bond, and type of element used

Types of bonding	Metallic	Ionic	Covalent	Intermolecular forces
Types of elements used				
Givers and/or takers				
Descriptions				
Types of materials formed				
Strength of bond				
Properties produced				

Table 2.2 Chart of what materials would be categorized under the heading of metals, ceramics, polymers, or composite. Everything is Materials, page 30 of solids, ASM Educational Foundation

Metals	Ceramics/glass	Polymers	Composites
Stainless steel foil	Glass stirring rod	Latex balloon	Mylar balloon
Metal scoopula	Fiberglass strands (no matrix)	Cotton cloth	Formica
Copper wire	Brick	FedEx envelope (Tyvek)	Cardboard
Aluminum wire	Calcite crystal	Preform	Plywood
Steel wool	Glass bottle	PVC pipe	Cut up ski
	Pumice	Acrylic cube	
	Talc	Rubber stopper	
	Shell[a]	Styrofoam cup	
	Glass foam block[b]	Candle[c]	

Additional Items:
Silicon lump – this doesn't fit into any of the categories. The only materials that are elemental are metals; the rest are compounds or mixtures. Silicon isn't a true metal. This item helps to teach the students that no classification system is perfect or will work for every possible sample. An additional category could have been added such as "semiconductor" or "electronic material"
Light bulb – this isn't a single material; it is an "assembly." It is made up of more than one material, and each material has its own function. The different materials are joined together but still are separate unlike a composite
[a]Shell – based on the type of shell, this could be listed under composites because the layers of ceramic are bonded by polymers. Accept it under either category and have a discussion with the students
[b]Glass foam block – used to clean grills or shave the "pills" off of sweaters
[c]Candle – the students will want to place it under composite because of the wick. Remove the wick if possible or tell them just to consider the wax. Some scientists would not classify the wax as a polymer because the chains are too short. This is a good point of discussion

Table 2.3 Material characteristics of a metal, ceramic, polymer, or composite distinguishing one from another. Everything is Materials, page 31 of solids, ASM Educational Foundation

Metals	Ceramics/glass	Polymers	Composites
Hard	Not conductors	Plastics (synthetic)	Combination of two or more material
Shiny (luster)	High melt T°	Some are natural	
Variable melt T°	Brittle	Some are absorbent	
Flexible	Some are transparent	Flexible	
Malleable	Fragile	Low melt T°	
Ductile	Breakable	Low density	
Good conductors	Not workable	Many are recyclable	
Heavy (dense)	Formed by heat	Stretchable	
Most are dense	Sand or clay based	Not brittle	
Opaque	Fairly high density	Bendable	
Undergo oxidation	Poor conductors	Rubbery	
Bendable	Can be formed	Expandable	
Strong	Includes minerals	Shrinkable	
Can be elements	Can be recycled	Elasticity	
Some are magnetic		Some are transparent	
Fairly high melt T°		Poor conductors	
Impermeable		Can be easily formed	
Workable		Long molecules	
Corrode easily		Some are hard	
Come from ores		Variety of properties	
Can be mixtures (alloys)		Made of compounds	

References

Dictionary.com (n.d.). Retrieved March 3, 2015, from http://dictionary.reference.com/

Sass, S. (2011). *The substance of civilization materials and human history from the stone age to the age of silicon*. New York: Arcade Pub. (Sass, 2011)

Chapter 3
Atomic Structure and Periodic Table

In the course of 150 years, the periodic table has evolved into the table of chemical elements we have today. This organization of elements, originating in 1800s, has given the scientific community a better understanding of the composition of elements. Electrons, protons, and the nucleus are terms taken for granted today because of the work done with the periodic table while in high school.

Atomic Structure and the Periodic Table

When investigating materials, the understanding of basic atomic structure is imperative. Beginning with elementary knowledge of protons, neutrons and electrons, their relative location, shell structure, and atomic mass, enables the scientist in their understanding of the composition of the material, its bonding structure, how can the material possibly be strengthen, as well as how the material can be altered to improve the original material against corrosion, stress and/or strain.

There are two chief purposes in all scientific endeavors. The one is to secure accurate knowledge. The other is to interrelate the various items of that knowledge through the progressively unifying principles. The degree of maturity of any science may be quite accurately judged by the extent to which the tested knowledge within its domain has been organized and integrated (Atomic structure and the periodic table, Robert L. Ebel, *J. Chem. Educ.*, 1938, 15 (12), p 575, https://doi.org/10.1021/ed015p575, Publication Date: December 1938).

Knowing the atomic structure assists the scientist in locating the element on the periodic table. The information on this periodic table is organized incorporating the number of protons, electrons, neutrons, and the atomic mass. Being familiar with the atomic structure and the information found on the periodic table, the scientist can easily determine the type of element given, its atomic mass, the number of protons and electrons, and the capacity of the number of electrons in the outer shell, the valence shell. Investigation in the periodic table and having knowledge of atomic structure are the most basic needs in the understanding of materials.

R. A. Wesolowski et al., *The World of Materials*,
https://doi.org/10.1007/978-3-030-17847-5_3

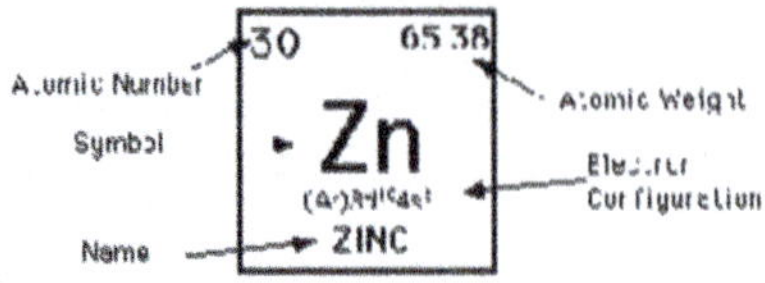

Fig. 3.1 Element from periodic table illustration to indicate what information is found on the periodic table for each element

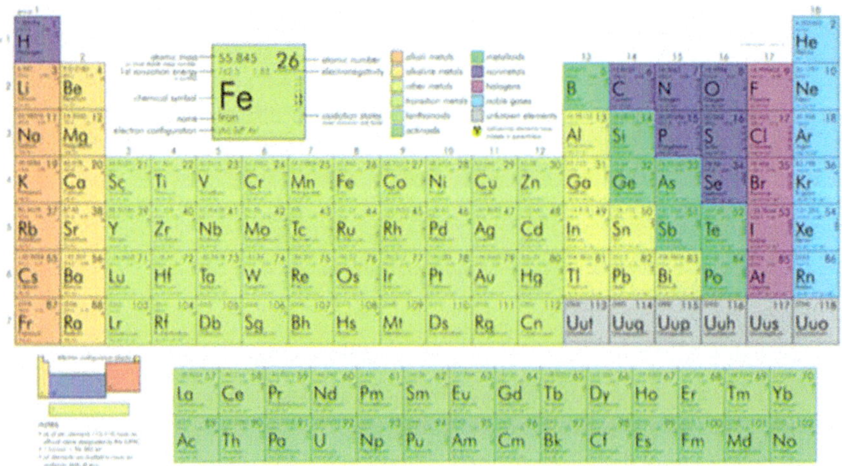

Fig. 3.2 Periodic table, File:Periodic table large.svg, From Wikimedia Commons, the free media repository

Organization and structure far exceed disorder and chaos. Placing elements in some organized pattern enables a structure to be used in finding commonality between existing elements. The organization of elements has been placed in a chart by a Russian chemist, Dmitri Mendeleev. His organization process was to place all elements with similar properties into a vertical column. In addition, those same elements were arranged by the atomic number likewise ordering the element's protons.

The rows of the periodic table are named "periods." They note the number of electron shells of an elements. The columns, called families or groups, are composed of elements with the same properties and the same number of valence electrons.

Recall that the valence electrons are the outmost shell of electrons and are involved in the bonding of the element.

The identification of the elements on the periodic table is consistent showing the atomic number, the mass, the element's configuration, the name, and the symbol for each element (Fig. 3.1).

Reference the periodic table below and determine which element is larger in size and justify your answer (Fig. 3.2).

Oxygen or selenium
Magnesium or sulfur

The magnesium element is larger than sulfur. Atoms tend to decrease in size moving across the period from the left to right. The nucleus has a greater pull toward the center of the orbit on the electrons, the further across the period (the horizontal row of the periodic table.)

Metals are located on the left portion of the periodic table. The metals have between one to three valence electrons. The metallic elements are "givers" of electrons. Having relinquished an electron, the element becomes positively charged. The nonmetals are located on the right side of the periodic table. They have five to eight electrons in their outer shell. Being rather stable, the nonmetals would serve as suitable recipients for additional electrons. This makes the element very negatively charged. The elements that are very negatively charged (anion) or are deficient of electrons and thus are positively charged (cation) are called ions. An ion is a charged particle which is not neutral.

Column 8 of the periodic table houses the noble gases. The column of noble gases is very stable, has a full valence shell, avoids bonding, is odorless, is colorless, and is monatomic gas.

The halogens are located in the seventh column on the periodic table. The valence shell is "s^2" "p^5," implying that the halogens desire additional electrons to complete the valence shell. The halogens are normally found as an ion in compound due to their high reactivity.

Column one contains the alkali metals. The alkali metals possess one electron in the valence shell. Alkali metals are highly reactive to water.

The alkaline earthen metals located in the second column ionize with the halogens to give off ionic salts. The earthen metals are very stable at high temperatures and having two electrons in the outer shell (Myers et al. 2006).

Not all materials are the same. Some materials are rigid; some materials are pliable. Some materials are shiny; some materials are dull. To describe materials by their properties an investigation can spotlight their chemical, physical, or mechanical properties.

Chemical properties or characteristics of the material can be characterized as combustible, chemically reactive, corrosion resistant, and toxic. The ability to catch on fire and burn is a description of a combustible property. Some materials have a propensity to react with other materials such as acids and bases, while others do not. Some materials are corrosion free, that is, resistance to oxidation. Finally consider materials that are high toxic, filling the air with fumes. Each of these characteristics has a chemical bases for an explanation. Materials do not react identically. Not all materials are affected by the environment in the same manner.

Physical characters do not change the chemical composition of matter. Discussing the traits, height, weight, color, or any other aspect inherent to the material is considered a physical description. Additional physical traits are melting point, boiling point, texture, solubility, luster, density, magnetism, odor, viscosity, and crystalline structure. A conductivity property may refer to heat and/or electricity. They possess the ability to either resist or conduct heat or electricity. When discussing heat, the heat capacity is referenced (i.e., the amount of heat needed to raise a temperature one degree). As heat is applied, materials expand; when cooled, it contracts. The rate of expansion or contraction, the amount of heat needed to raise a temperature, is unique to a particular material. The optical characteristic or a material simply implies that the material is opaque, translucent, or transparent. Does it have a particular refraction (bending of a wave) characteristic? translucent or transparent. Does it have a particular refraction (bending of a wave) characteristic?

Mechanical characteristics, a subgroup of the physical group, involve an application of a stress or a strain (application of a force), brittleness, hardness, elasticity (ability to stretch and return to an original shape) or plasticity (going beyond the elastic state, a deformity occurs), and workability (where material can be malleable (flattened) or ductile pulled). Toughness is the ability to resist a fracture or absorb energy. Strength is a resistance to distortion due to a stress or a force. Types of strengths are tensile strength (pulling on a trash bag), a compression (pressing an object together), torsion (twisting an object), bending, and shearing (cutting with a pair of scissors.)

Reference

Myers, R., Oldham, K., & Holt, R. (2006). *Holt chemistry*. Orlando: Holt, Rinehart and Winston. (Meyer, 2006)

Chapter 4
Crystal Properties

Most of chemistry focuses on gases or liquids, and occasionally solids are introduced. The solid material is basically the materials most individuals use. Metals, ceramics, and plastics and combination of these materials make up many of substances in the world. To investigate the basic structure of the solid, an investigation in crystal structure is essential.

Crystal Models

Crystal structure is an orderly arrangement of atoms, ions, or molecular materials in a crystalline (orderly structured) material. The created structure is in three-dimensional space. The crystal structure is composed of a *unit cell*, a set of atoms constructed in a particular way. The crystal structure and symmetry play a role in determining many physical properties, such as cleavage (a sharp division; a split), electronic band structure (the range of energies that an electron within the solid may have (called energy bands, allowed bands, or simply bands) and ranges of energy, and optical transparency (the physical property of allowing light to pass through the material without being scattered).

"Crystal structure is one of the most important aspects of materials science and engineering as many properties of materials depend on their crystal structures" ("crystal structure"). Crystals are a study in solid-state physics. "If we postulate that crystals are made of congruent building blocks stacked one upon the other, one next to the other, each of the laws can be derived" (Langan 1989). Crystals are made of atoms, and the atom's formation is based on electrons, protons, and neutrons contained in the atom. The arrangement of the atom helps define the crystal structure. Atoms may be structured in two ways: they may be organized, well-structured, crystalline, or they may be structurally chaotic, lacking structure, and amorphous. The structure of a crystal will help identify some of its material properties. Investigating the crystal structure of a solid, where the atomic structure is fixed in space, as opposed to being structured in a fluid state as found in liquids and gases, permits an investigation of properties in that solid. A crystalline solid has a

R. A. Wesolowski et al., *The World of Materials*,
https://doi.org/10.1007/978-3-030-17847-5_4

well-defined melting point. Since the distance between atoms is the same due to the regularity of the crystalline shape, and the created environment around the atom is the same, the intermolecular force holding the atoms together is the same due to the ordered structure of the crystal. Because of this similar atomic arrangement, the amount of thermal energy needed to break the bonded atoms would have to be the exact amount for each atom and would need to be distributed throughout the crystalline structure at the same time. An amorphous structure is not organized; rather it consists of poorly defined patterns. Subsequently, the energy needed to break the atomic structure would vary throughout the crystal structure.

There are other types of crystalline solids: ionic solids, molecular solids, covalent-network (also called atomic) solids, and metallic solids. The ionic solid is held together by electrostatic force. The solid is brittle, possessing a high melting point and is a poor conductor. An example of the ionic solid would be table salt (NaCl). The molecular solid has atoms held together by dipole-dipole forces, hydrogen forces, or London forces. "Dipole-Dipole interactions result when two polar molecules approach each other in space. When this occurs, the partially negative portion of one of the polar molecules is attracted to the partially positive portion of the second polar molecule. This type of interaction between molecules accounts for many physically and biologically significant phenomena such as the elevated boiling point of water" (Johns and Guess 2014). "A hydrogen bond is a weak type of force that forms a special type of dipole-dipole attraction which occurs when a hydrogen atom bonded to a strongly electronegative atom exists in the vicinity of another electronegative atom with a lone pair of electrons. These bonds are generally stronger than ordinary dipole-dipole and dispersion forces, but weaker than true covalent and ionic bonds" (Pietri 2014). "The attractive forces that exist between molecules are known as intermolecular forces. These include ionic interactions, dipole-dipole interactions and dispersion or London dispersion forces. Dipole-dipole interactions and dispersion forces are weaker than thermal energy (2.4 kJ/mole) at room temperature and are referred to as Van der Waals Force. The London dispersion force is the weakest intermolecular force. It is a temporary attractive force that results when the electrons in two adjacent atoms occupy positions that make the atoms form temporary dipoles" (Ilagan 2014). The solid is flexible, possessing a low melting point and is a poor conductors of electron flow. Sucrose is an example of the molecular solid. The covalent-network or the atomic solid is made up of covalent bonds. Intermolecular bonds are also covalent bonds. Characteristic of an atomic solid is that it processes a very high melting point but is a weak conductor. Diamonds and graphite are examples of this covalent-network bond. The metallic solid possesses atoms held together with metallic bonds. The characteristics of a metallic solid are that it possesses high melting points, is an excellent conductor of electricity, and may range from a soft malleable material to a very hard solid.

The crystal structure is a three-dimensional structure created by the stacking of two-dimensional structures on top of each other. One three-dimensional segment of the layering of atoms is a representation of a unit cell (Fig. 4.1). A unit cell is the simplest form of the atomic configuration that can be repeated to create the crystalline solid. The unit cell would be represented in far right illustration above, and the

Fig. 4.1 Simple crystal

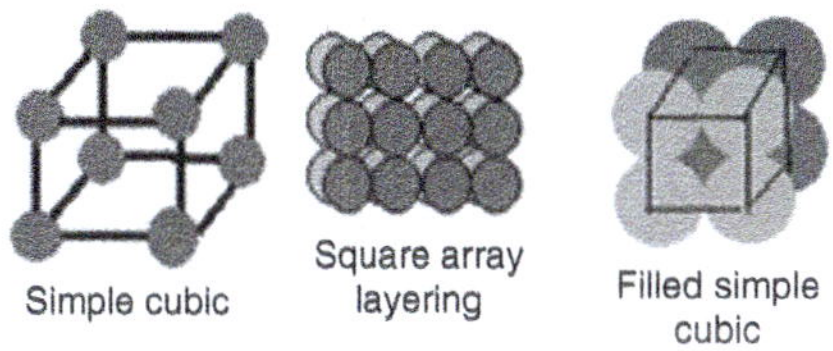

http://www.chem.fsu.edu/chemlab/chm1046course/solids.html

Fig. 4.2 Volume of simple cubic, illustrating the relations between volume of the atoms and the volume of the cube. Self-drawn, R Wesolowski, 2015

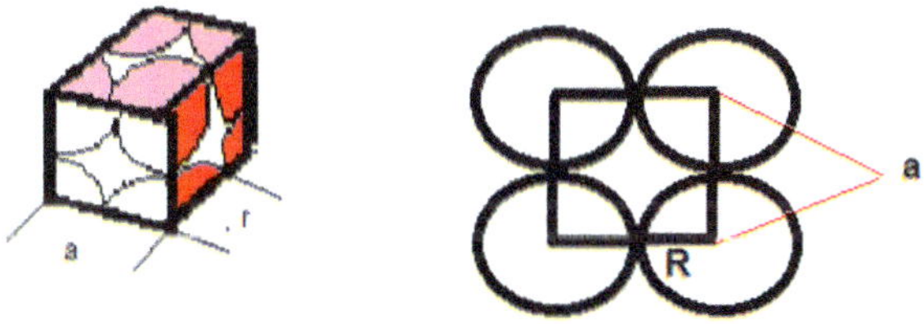

center figure represents the crystalline solid. Placing a planar configuration, a two-dimensional model, directly over the initial crystal structure, creates a basic "simple cubic structure."

Initially consider the volume of a simple cubic structure and determine the atomic packing factor, the percentage of the volume occupied by the atoms. Referencing to the above Fig. 4.1 Filled Simple Cubic, the simple cubic is made up of eight corners with each corner consisting of one-eighth of an atom. The total number of atoms in the simple cubic is one. The volume of an atom is $4/3\pi r^3$ (Fig. 4.2).

The length of two radii would equal the length a and $r = a/2$. The volume of the cubic would be a^3, and the volume of the atoms within the cubic would be

$$4/3\pi r^3 = 4/3\pi\left(a/2\right)^3$$

$$\text{Volume} = 0.52\,a^3$$

Having this volume of 0.52 a^3 implies that 52% of the solid cubic is occupied by the volume of the atom of a simple cubic. In other words, 48% of the cubic is not occupied.

Metals and most solids can be characterized as body-centered, face-centered, or hexagonal-centered cubic structure. The body-centered cubic (BCC) has an atom located at the four corners of the cubic structure similar to the simple cubic and an additional atom located in the center of the simple cubic configuration. The BCC structure is composed of two atoms. There are eight corners, and each corner possess one-eighth of an atom per simple cubic structure. Additionally there is one atom located in the center of the configuration (Fig. 4.3).

Fig. 4.3 Body center cubic, ice VII structure (aka body-centered cubic lattice, bcc) by Vitroid, published May 29, 2014

Lithium, sodium, potassium, chromium, barium, vanadium, alpha iron, and tungsten are a few examples of materials that are body-centered cubic. Materials that are BCC are harder and less malleable than materials that are more closely packed. Deformed materials happen as the planes of the atoms "slip" over each other. This is difficult in a body-centered cubic structure. The spacing between the atoms in a body-centered cubic structure allows for impurities to enter into the crystal model filling the void, retarding the potential of slipping, and thus preventing the deformation. The hardening procedures will be explained later.

The face-centered cubic is similar to the body-centered cubic in the simple cubic structure having eight corners possessing one eight of an atom per corner. The difference between FCC and BCC is that the FCC does not have an atom in the center of the structure; it has a fractional part of an atom on each face of the atom for a total of three additional atoms (Fig. 4.4).

The atoms in the FCC are packed more closely together than in the BCC. Observe the way the atoms are placed on top of each other. Having an atom on each face, the second row of atoms would shift to the right or left placing one half of an atom in the center of the simple cubic structure that it rests. This configuration reduces that unoccupied space between the atoms. With the reduction of the space between the atoms, there is less room for impurities to enter into the atomic structure. The face-centered cubic will have a propensity to be more ductile, having an ability to deform without losing strength. As the BCC is not as closely packed as the FCC, the BCC forms a stronger less ductile material. The ductility of material which defines a characteristic of the metal is more important than the crystal structure, and thus awareness of the atomic packing factor is critical. This too will be discussed later. Aluminum, copper, gold, iridium, lead, nickel, platinum, and silver are FCC-structured materials.

A hexagonal closed packed crystal is a structure where the layering of the crystal is distributed over three layers as illustrated below (Fig. 4.5). The top layer has six atoms arranged in a hexagonal shape and one additional atom in the center, and the center layer has three atoms in a triangular shape resting itself between the space between two atoms (Fig. 4.6). The hexagonal structure is common for elemental metals. Some examples of the HCP structure are beryllium, cadmium, magnesium, titanium, zinc, and zirconium.

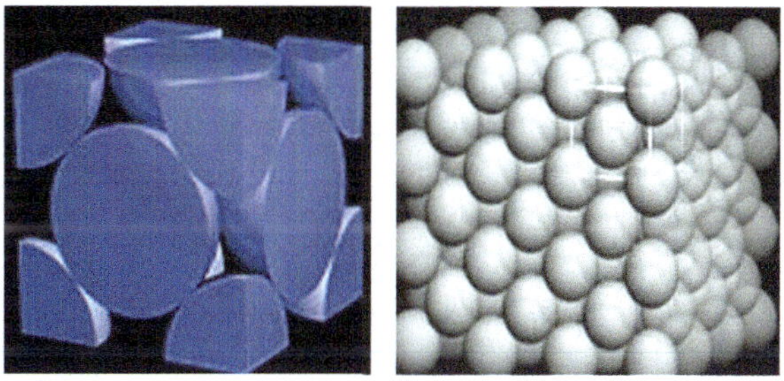

Fig. 4.4 Face-centered cubic, Primary Metallic Crystalline Structures www.nde-ed.org 598 × 225 Search by image, "In the FCC structure (and the HCP structure) the atoms can pack closer together than they can in BCC structure"

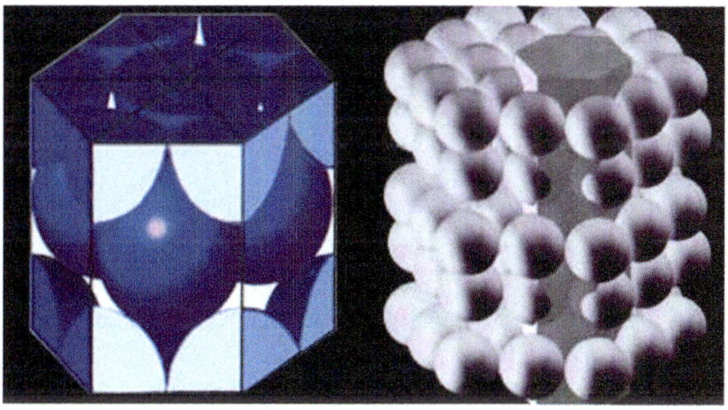

Fig. 4.5 Hexagonal cube, Primary Metallic Crystalline Structures, www.nde-ed.org 598 × 237 Search by image. "In each the top and bottom layer, there are six atoms that arrange themselves in the shape of a hexagon and a seventh atom that sits in the middle of the hexagon. The middle layer has three atoms nestle in the triangular 'grooves' of the top and bottom plane. Note that there are six of these 'grooves' surrounding each atom in the hexagonal plane, but only three of them can be filled by atoms"

Fig. 4.6 Hexogonal cubic construction, Everything is Materials, page 68 of solids, ASM Educational Foundation

The atomic packing factor has been referenced previously. Consider placing a cubic structure into a cubed container. Had the container been filled with water to the extreme top prior to placing the cubic structure into the container, an amount of water would be displaced. This displaced water would be the volume of the crystal. The remaining water not displaced would be the amount of volume of the void found within the crystal structure (Fig. 4.7).

Determine the percentage of space occupied in the *BCC* and the amount of void in the *BCC*. We will define the radius *R*, in terms of the side of the cube to establish

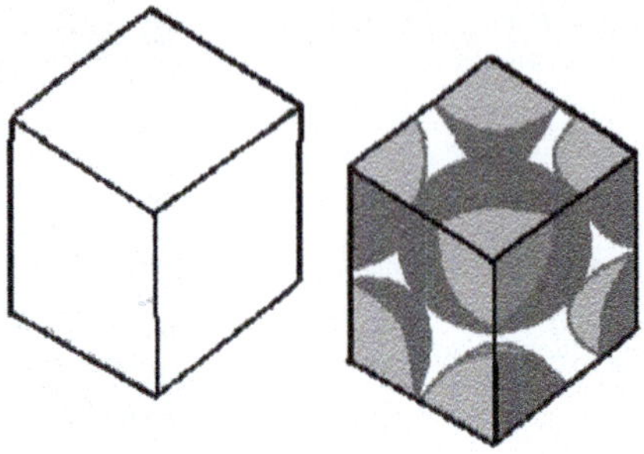

Fig. 4.7 BCC crystal structure, illustrating the volume comparison of the cubic and the crystal

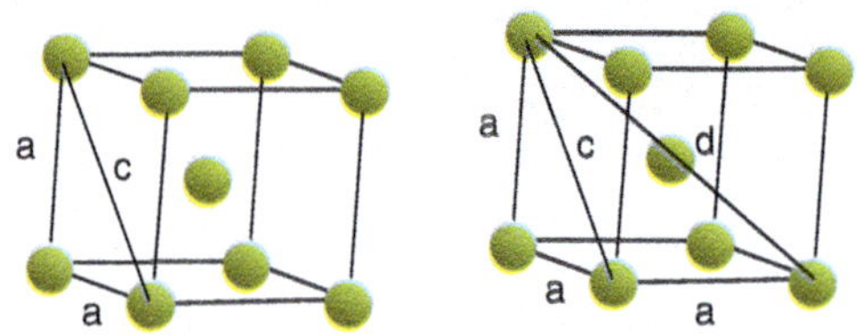

Fig. 4.8 BCC cube, illustrating the diagonal along the face and the diagonal across the interior of the cube. (**a**) BCC cube; (**b**) BCC cubic

a common unit relating the volume of the atom to the volume of the container. We can then determine the volume of the cube in terms of *R* and then the volume of the cubes (Fig. 4.8).

Using the Pythagorean theorem, the diagonal of the side *c* is found by solving

$$c^2 = a^2 + a^2 \; ..... \text{ then } c = \sqrt{2}a$$

We can now solve for the diagonal *d*, using *c* and the base *a*.

$$\text{substituting} \quad d^2 = c^2 + a^2,$$

$$\text{then} \quad d^2 = \left[\sqrt{2}a\right]^2 + a^2$$

$$d = \sqrt{3}a$$

Consider the diagonal… the cube is a creation of atoms measure center to center… the diagonal is a measurement of the radius *R* of ½ an atom, at the one vertex point, the cube in the middle, having **2 R**, and another ½ atom at the other vertex point for a value of **R**. Thus the diagonal would have a length of **4 R** and so, $\mathbf{d = \sqrt{3}\,a = 4R}$, and then solving for **a**, we have… $\mathbf{a = 4\,R/\sqrt{3}}$. The volume of a cube of side **a**, is $\mathbf{a^3}$, since $\mathbf{a = 4\,R/\sqrt{3}}$, then $\mathbf{a^3 = 64\,R^3/3\sqrt{3}}$.

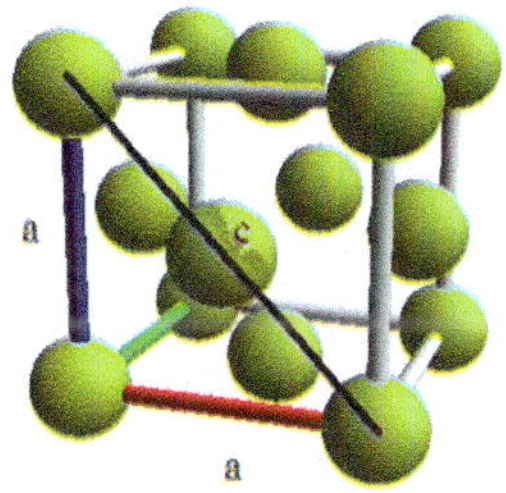

Fig. 4.9 Face-centered cubic, illustrating the relationship between the diagonal along the face of an FCC and the length of the cube

The volume of a cube is **4/3 πR³**... Each vertex point would be made up of 1/8 of a cube. There are 8 such vertex points so we have one volume of **(4/3) πR³**, but this is body-centered cubic, and there is a cube in the middle in addition to the vertex points, so the total volume of the atoms would be **2 × [(4/3)π R³]** or **(8/3)π R³**. The percentage of occupied space would be.... **[8/3π R³]/[64 R³/3√3]** = **√3 π/8** or **68%** filled; this implies there is a **32%** void.

Determine the percentage of space occupied in the BCC and the amount of void in the BCC. Initially we will define the radius *R*, in terms of the cube. We can then determine the volume of the cube in terms of R and then the volume of the cubes (Fig. 4.9).

The diagonal *c* can be evaluated using the Pythagorean theorem...$c^2 = a^2 + a^2$ or c is equal to $\sqrt{2}\,a$. Consider the diagonal... the cube is a creation of atom's measure center to center... the diagonal is a measurement of the radius *R* of ½ an atom, at the one vertex point, the cube in the middle, having 2*R*, and another ½ atom at the other vertex point for a value of *R*. Thus the diagonal would have a length of 4*R* and so, $c = \sqrt{2}\,a = 4R$ and $a = 4R/\sqrt{2}$.

The volume of the cube would be a^3 which would equal to **[4R/√2]³ = 64 R³/2√2**. The FCC would have 8 vertex points each being 1/8 the value of a cubic volume of a sphere **[(4/3)π R³]**. Each face has ½ a cube in the center. There are six centers so this would make three addition spears so the total volume of the cubes would be **4 × [(4/3)π R³] = (16/3)π R³**. The percentage of occupied space would be **[(16/3)π R³ /64 R³/2√2]** or **(2√2)π/12 ~ 0.74 or 74%**; the amount of void is **26%**.

In summary, atoms are assembled into an amorphous or crystalline structure. The common structures for metals are body-centered cubic (BCC), face-centered cubic (FCC), and hexagonal closed packed (HCP). The atomic packing factor is the same for FCC and HCP, and both the FCC and HCP atomic packing numbers are greater than the BCC. Prediction of the density of a material may be determined provided there is awareness of the atomic weight, atomic radius, and crystal geometry (e.g., FCC, BCC, HCP).

The crystal properties of materials are dependent on the crystal structure of the material. Understanding these properties and the effects of the properties on materials is a major concern to a material scientist and engineer.

References

Ilagan, R. (2014). Chemistry LibreText, *London Dispersion Interactions*, https://chem.libretexts.org/Bookshelves/Physical_and_Theoretical_Chemistry_Textbook_Maps/Supplemental_Modules_(Physical_and_Theoretical_Chemistry)/Physical_Properties_of_Matter/Atomic_and_Molecular_Properties/Intermolecular_Forces/Specific_Interactions/London_Dispersion_Interactions

Johns, D, Guess, J. (2014). Chemistry LibreText, *Dipole-dipole interaction*, https://chem.libretexts.org/Courses/University_of_Georgia/CHEM_3212/01%3A_The_Properties_of_Gases/1.09%3A_Specific_Interactions/1.9.01%3A_Dipole-Dipole_Interactions

Langan, FJ. (1989). Crystals: Geometry and Groups, Curricular Resources. Volume VI, Unit 5 (89.06.05), Section 2

Pietri, J. (2014). Chemistry LibreText, *Hydrogen Bonding*, https://chem.libretexts.org/Bookshelves/Physical_and_Theoretical_Chemistry_Textbook_Maps/Supplemental_Modules_(Physical_and_Theoretical_Chemistry)/Physical_Properties_of_Matter/Atomic_and_Molecular_Properties/Intermolecular_Forces/Specific_Interactions/Hydrogen_Bonding

Chapter 5
Crystalline and Amorphous Solids

Energy, energy, everywhere.... In the investigation of the structure of solids, the crystal structure may be organized or disorganized. The characteristics of the material, malleability and density, are a result of the arrangement of the crystal structure.

Crystalline and Amorphous

When discussing crystalline and amorphous, a reference in the crystal structure is made. Crystalline refers to the orderly structure of the atoms, while the amorphous crystals are disorderly. This difference refers to the difference in the terms, but if a metal is crystalline in structure and another metal is amorphous in structure how does this difference manifest itself in the solid? Is it stronger or weaker? More apt to be brittle or is it stronger and harder? Observing the difference is keen; however the importance is to know how the difference affects the structure of the metal.

To investigate this difference, work in the mechanical, physical, or chemical properties can serve to better understand the material properties of the metals.

Crystalline solids are a class of solids that have atoms arranged in an orderly structured manner. Crystalline solids appear in four varieties: molecular, covalent, ionic, and metallic.

Molecular crystals are held together by a weak force known as Van der Waals forces. Van der Waals forces are forces that occur when molecules collide or pass near another molecule. It is a weak force. This force is the result of a rich region of electrons entering the sparsely populated area of electrons. The Van der Waals force is an intermolecular force causing atoms to adhere in liquid and solid states of other atoms of matter. The force is responsible for surface tension and capillary action (Dictionary.com 2015). Common examples of this mixture of molecules are dry ice, wax, and iodine. Their molecular crystal is soft and compressible, having a low melting point and boiling point. They are poor conductors of electricity but are considered as good electrical insulators.

R. A. Wesolowski et al., *The World of Materials*,
https://doi.org/10.1007/978-3-030-17847-5_5

Ionic crystals, crystals that are bound together by electrostatic attraction (N.A. 25 November 2014), are atoms attracted to other atoms through the positive and negative charged ions arranged in an orderly fashion throughout the crystal. Due to their oppose charges, the attractive force is a strong electrostatic force. Common examples of ionic crystals would be table salt (NaCl). Hardness and brittleness are characteristics of an ionic crystal. The material used in ionic crystals is a poor conductor of heat and electricity. Instead, ionic crystals are used as a resistor in solid-state circuits. Ionic crystals have a high level of vaporization; consequently, they possess low vapor pressure. In addition to having a high enthalpy of fusion (i.e., transitioning from a solid to a liquid (Dictionary.com 2015)), when the ionic crystal is dissolved in polar solvents (i.e., a liquid with molecules that have a slight electrical charge due to its shape (WiseGeek.com 2015)), these crystals conduct electricity.

Covalent crystals are created through a bonding of two or more constituents. This bonding may be the result of two similar or dissimilar materials. Examples of the covalent crystal are diamonds and quartz. A covalent crystal is incompressible and hard. The material possesses a very high melting point and is a very poor conductor of electricity at any temperature. Covalent crystals have a high heat of fusion (i.e., the heat absorbed of a given solid at its melting point completely equal to the heat of solidification (Dictionary.com)) and a high enthalpy of fusion (i.e., the sum of its internal energy and the product of its pressure and volume (Dictionary.com)), during the transitioning from solids to liquids.

Metallic crystals have positive kernels located within the nuclei, where the inner electrons are dispersed in a sea of mobile valence electrons (Bonding in Solids n.d.). The forces created due to moving electrons create a metallic bond. Examples of metallic crystals are common metals such as copper, nickel, and alloys. The metallic crystals may be hard or soft. They are excellent conductors of electricity and heat. The material has high luster and reflectivity. These crystals may be malleable and ductile (able to be hammered or pulled.) Metallic crystal may have moderate heats of fusion (transitioning from solid to liquid.) This unique characteristic of metallic crystal enables a great variance to coincide in the properties of different metals.

What are amorphous metals? As stated previously, crystalline metals are orderly structured to the atomic formation. An amorphous solid is not arranged in any organized lattice pattern. Glass, gels, or plastics fall into this category. The amorphous material is created by quickly cooling the melted material or by quenching. This rapid cooling prevents the mobility of the crystals from rearranging itself in an organized structured pattern. Another method of creating an amorphous solid is by adding impurities to a material inhibiting crystallization. An example of the impurity addition is adding soda to silicon dioxide which renders window glass. Two characteristics of an amorphous solid are that they have a higher strength than a crystalline solid and that they possess better corrosion characteristics.

An amorphous material, glass, is created due to a rapid cooling prior to crystallization. Glass becomes brittle and often optically transparent. Polymers may be amorphous, crystalline, or even semicrystalline. Glass transition temperature T_g is the temperature where the mechanical properties of the glass making material (silica found in sand) radically change due to internal movement liquefy and then transition into a solid. If a polymer experiences temperatures above the transition temperature of glass, the polymer becomes elastic. However, if the temperature is below the transition temperature, the amorphous solid becomes brittle and hard.

The word defect often connotes negative inferences. However, in materials science, a defect referred to as a discontinuity whose size, shape, orientation, or location makes it detrimental to the useful service of the part that it occurs (Reardon 2011) may enhance a product, strengthening it or even formulating a superior product. A defect can be a good or a bad thing. Impurities or imperfections of the atoms in a material affect the physical properties of a material. There are three primary types of defects: point, line, and interfacial (Fig. 5.1).

In a vacancy defect, an atom is missing in the lattice structure. This defect can move around the atom through diffusion causing an intermingling of molecules, ions, etc., resulting from random thermal agitation, as in the dispersion of a vapor in air (Dictionary.com 2015) (Fig. 5.2).

An interstitial defect is placing atom in the spacing where no atom would be located. This creates a high-energy configuration. This defect is a technique used in creating alloys (Fig. 5.3).

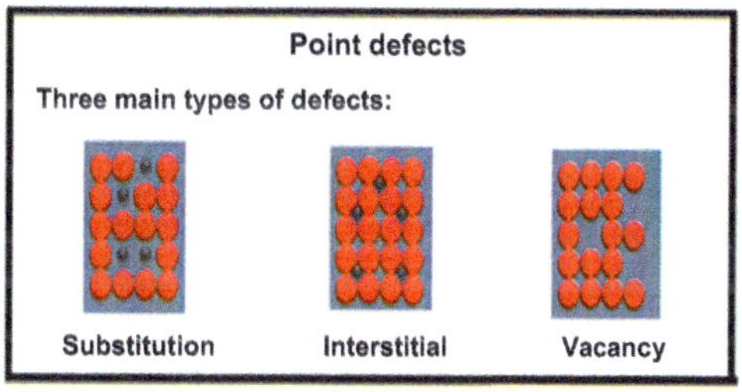

Fig. 5.1 Point defects, illustrating the types of point defects

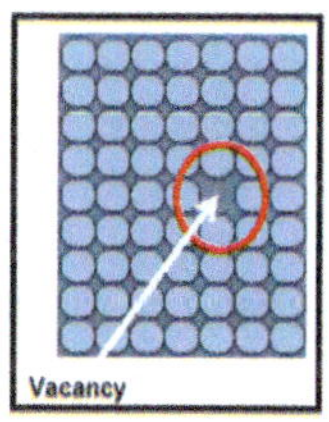

Fig. 5.2 Vacancy, illustrating a vacancy in a point defect

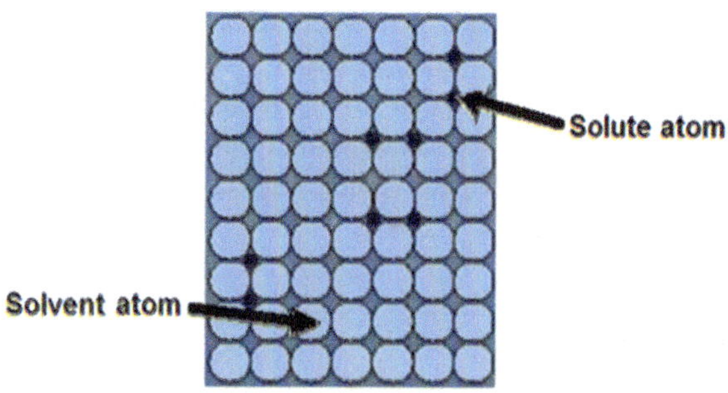

Fig. 5.3 Interstitical defects, (na) (2012) Classifications of Materials, Cleveland, Ohio, ASM Material Education Foundation

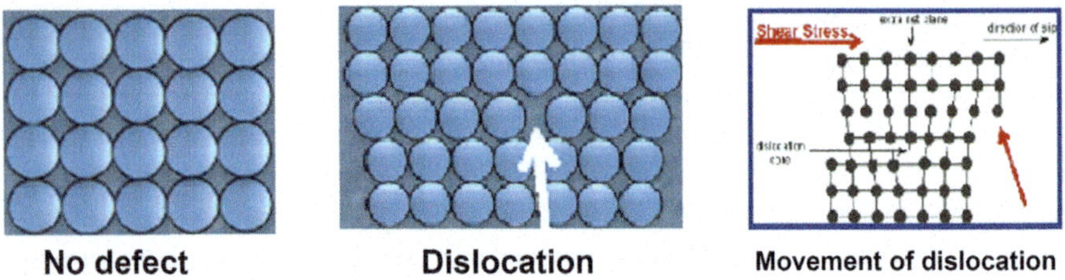

Fig. 5.4 Defects: dislocations and movements of dislocations, showing movement of dislocations

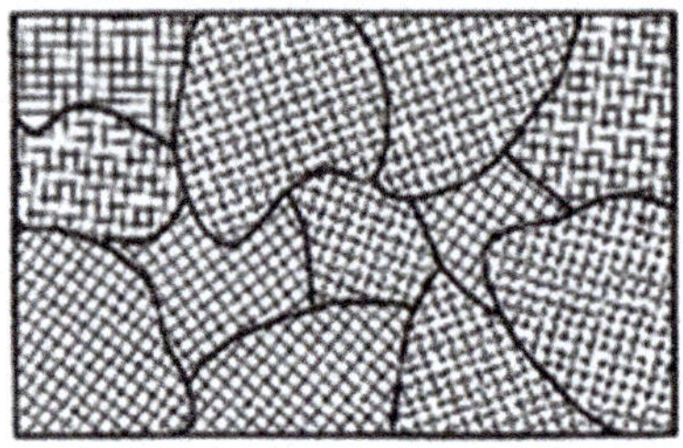

Fig. 5.5 Grain boundaries

Substitution defect occurs when one or more of the original atoms are replaced by different atoms. This process is used to create an alloy.

A dislocation is a line defect where the region has atoms that are not perfectly aligned. This defect can move. Where many such defects exist, the metal is hard. When there are fewer defects, the metal is more workable. An application of a stress to a line defect can result in the entire defect to shift, creating a step formation (Fig. 5.4).

As stated in earlier material, a grain boundary is an interface where grains of a polycrystalline material meet. This grain boundary is a planar defect. This planar defect disrupts the motion of dislocations through a material. A disruption of motion reduces the grain size but enhances the strength of the crystal (Fig. 5.5).

The manner in which the grain is prepared affects the grain size. If a grain is quickly cooled, the result is a small grain. When the grain is slowly cooled, the result is a larger grain. There are more grain boundaries in a smaller grain than in a larger grain. At low temperatures, the smaller grain produces more grain boundaries which results in an increase of strength in the material. The grain boundary is a barrier to dislocated movement. At high temperatures, the opposite occurs in that the smaller grains reduce their strength because grain boundary movement may occur. In these cases, the grain boundaries become a source of weakness.

Grain boundaries can affect other material properties. Changing the heat can alter the electrical and/or thermal conductivity and improve material strength by reducing dislocated movement, and an onset of corrosion may ensue when high energy is in effect.

Movement of atoms in a material is referred to as diffusion. This movement is paramount in the heat treatment of materials, the manufacturing of ceramics, and/or the solidification of materials. It must be stressed that in the creation of materials, a multitude of atoms are introduced. In the construction of an engine, there is a combination of ceramics with metals. The ceramic adds high temperature stability, while metal adds strength and durability. The material scientist examines whether the atoms will diffuse into each other or will they remain separate entities. Owing to the tight atomic structure of solids, motion is slow and constant.

There are various types of diffusions (i.e., migration or a movement of an intermingling of molecules, ions, etc., resulting from random thermal agitation (Dictionary.com)). A self-diffusion is the movement of atoms through their lattice structure. Interdiffusion, impurity diffusion, is the movement of particles that are not part of the crystal structure. The diffusion occurs within a vacancy defect or an interstitial defect. To perform the movement of the atom, the atom needs an empty adjacent site while retaining enough energy to break bonds and cause lattice distortion during displacement.

In vacancy diffusion, the atom moves its position from a normal lattice positon to an atomic void within the lattice structure. The interstitial diffusion has atoms moving within the voids or spaces of the atomic structure. The diffusion atoms are small compared to the atoms of the crystal. The interstitial diffusion occurs more rapidly than the vacancy diffusion. When the diffusion process is initiated, a great deal of energy is required to have an atom squeeze past a neighboring atom into a vacancy. Activation energy (i.e., the least amount of energy required to activate atoms or molecules to a state in which they can undergo a chemical reaction (Dictionary.com) is the energy required to force the atom to a new location). The number of vacancies within the crystal lattice can be calculated at a particular temperature, and the rate of diffusion can be determined using Arrhenius rate equation. Both of these procedures are reserved for a higher level of materials science. The discussion of diffusion is important because it is the procedure used in the making of steel and the introduction of carbon into the iron lattice, and impurity diffusion is applied in the creation of a silicon crystal.

Investigating metals requires a journey throughout history, focusing on the physical and chemical changes of the metallic composition. Examine the changes in history beginning with the Stone Age where the use of stone in its rudimentary form was used as the spear heads or the carving device, to the Bronze Age where the process of smelting of copper and tin created from the ore making bronze tools. Iron was used to make tools, hence the term Iron Age. However, due to poor refining processes, the metal was too soft. Eventually the inclusion of an impurity (carbon) through the heating process resulted into the making of steel. Reflect on the varying and significant properties of the metals throughout history. Where any particular metals would be found was dependent on the geographic location of people at a given time? (Fig. 5.6)

Consider the amount of metals that are available on the earth (Fig. 5.7).

Metals are found on the earth as minerals, and up to the seventeenth century, only 12 known metals were known to exist. Minerals were pulled from the earth and processes. Metallurgy, the technique or science of separating metals from their ore, (Dictionary.com 2015), is one of the oldest applied sciences. Some of the common metals are, iron, aluminum, copper, titanium, and nickel. Notice their position in the periodic table (Fig. 5.8).

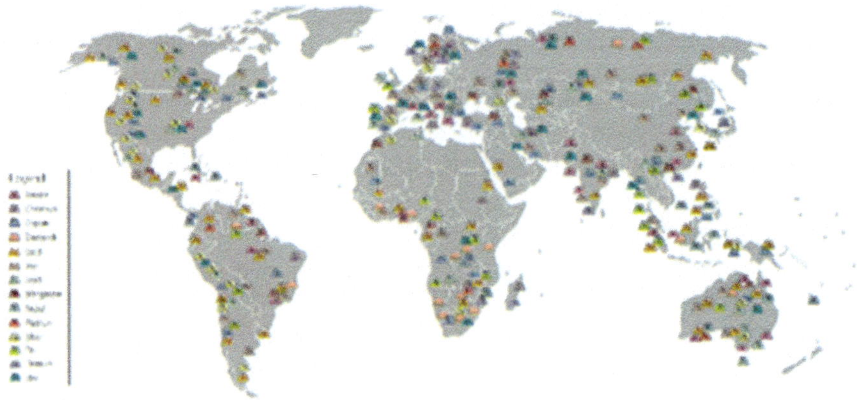

Fig. 5.6 Map of mining locations around the world, http://www.mapsofworld.com/world-mineral-map.htm

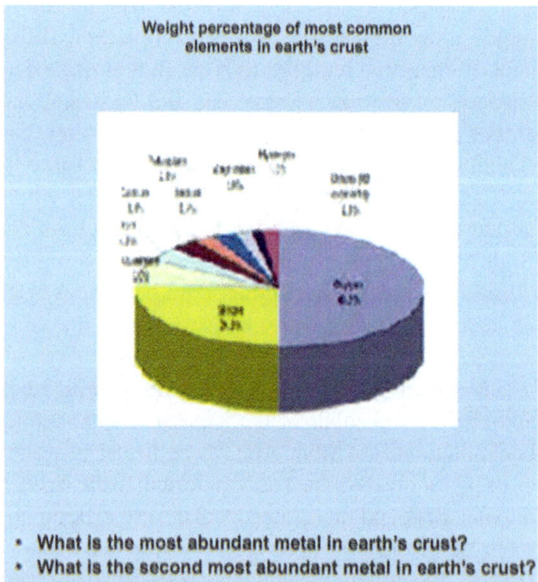

Fig. 5.7 Availability of materials in the earth, www2.latech.edu

These metals have low electronegativity, a measure of the ability of a specified atom to attract electrons in a molecule (Dictionary.com 2015). The valence electrons, electrons located in the outer most shell, are delocalized and create a sea where they move around freely. Characteristics of metals are described as physical when we speak of the luster or density. They are described as mechanical when referencing the malleability or ductility. Finally, they are referred to as electrical when referencing the conductivity, electrical, or thermal.

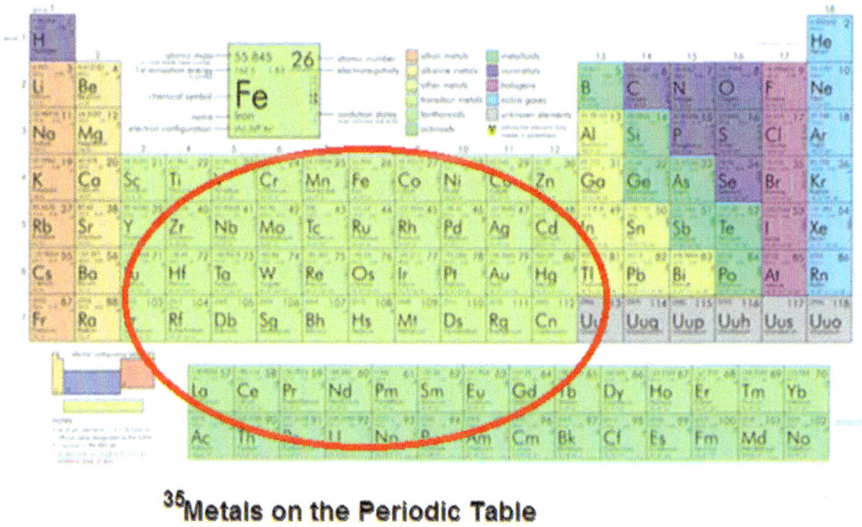

Fig. 5.8 Metals on the periodic table, Periodic Table, File:Periodic table large.svg. From Wikimedia Commons, the free media repository

Table 5.1 d-Transition metals

Sc	Ti	V	Cr	Mn	Fe	Co	Ni	Cu	Zn
Y	Zr	Nb	Mo	Tc	Ru	Rh	Pd	Ag	Cd
La	Hf	Ta	W	Re	Os	Ir	Pt	Au	Hg
Ac	Rf	Ha	Sg	Ns	Hs	Mt			

The metals are grouped according to commonality among the metals. The first group, the alkali metals, is located in the first column of elements of the periodic table, beginning with hydrogen, the first element. These metals share common features. They all possess a low melting point, only have one valence electron, possess a lower density than other metals, low electronegativity, and ionization, and are a soft shiny metal.

The second column of metals is the alkali earth metals. These metals are silver in color and are also soft. This group of metals reacts with the halogens and water. Their reaction to water is not as severe as the alkali metals. The transitional metals are located in the center of the periodic table (Table 5.1).

These metals have an incomplete d-shell, have stronger bonds, and have higher melting points and boiling points. These sets of metals are the metalloids or semi-metals. These alloys are a transition between metals and nonmetals, and they are an excellent semiconductor. A sample of the metalloids is silicon which has a silver luster but is an inefficient conductor and is brittle. A history of metals, a paper written (Appendix A) by Dr. Alan Cramb, is located in the back of the text.

References

Bonding in Solids. (n.d.). Retrieved February 1, 2015, from http://www.mikeblaber.org/oldwine/chm1045/notes/Forces/Bonding/Forces08.htm

Dictionary.com (n.d.). Retrieved March 3, 2015, from http://dictionary.reference.com/

N.A., 25 November 2014 Source: Boundless. "Ionic Crystals. "*Boundless Chemistry*" Retrieved January 25, 2015 from https://www.boundless.com/chemistry/yexybooks/boundless-chemistry-textbook/liquids-and-solids-11/typers-of -crystals-87/ionic-crystals-383-6876/: (Boundless. "Ionic Crystals. "*Boundless*)

Reardon, A. (2011). *Metallurgy for the Non-Metallurgist, 2nd Ed.* A S M International. (Reardon, 2011

WiseGeek.com. (2015). http://www2.ucdsb.on.ca/tiss/stretton/CHEM2/arch20.htm

Chapter 6
Mechanical Properties

How can a force be applied to an object and what affect does the application of a force have on that object? From the basic definition of a force, a push, or a pull, we can expand this definition to stresses, strains, twisting, or cutting and how a particular material is affected by the force application.

Properties of material refer to those characteristics which describe matter. Chemical, physical, mechanical, and thermal properties are used to identify and describe a material. A mechanical property refers to matter which responds to force or strain. The response referenced to is how the material responds to force (push or a pull) or strain (force causing a distortion or deformation, force per area). Types of forces or stress used to investigate a material are tension (a pulling force that stretches a material), compression (a pushing force squashing an object), torsion (a twisting force), and shear force (opposing force or cutting force) (Fig. 6.1).

Response refers to the manner in which materials react to force (push or a pull) or strain (force causing a distortion or deformation, force per area). The modes of forces or stress used to investigate a material are tension (a pulling force that stretches a material), compression (a pushing force squashing an object), torsion (a twisting force), and shear force (opposing force or cutting force) (Fig. 6.2).

Metal properties alter under specific condition. With the addition of dissimilar material to metal properties known as alloying, the new composite is altered. Cooling an object at varying rates, quenching (i.e., quickly submerging into a cold water bath) or annealing (i.e., gradual cooling with air) may result in the material becoming brittle. The atomic structure of a material may also be altered through the process of work hardening. Consider a copper wire; twist the copper repeatedly. The atomic structure is being restructured. With continued twisting of the wire, heat is being transferred. At the same time, the ability to continue twisting the wire becomes more difficult. The integrity of metallic material may have been altered, rendering it defective. A defect is an imperfection in a material. The defect may be a void in the atomic structure (Fig. 6.3).

R. A. Wesolowski et al., *The World of Materials*,
https://doi.org/10.1007/978-3-030-17847-5_6

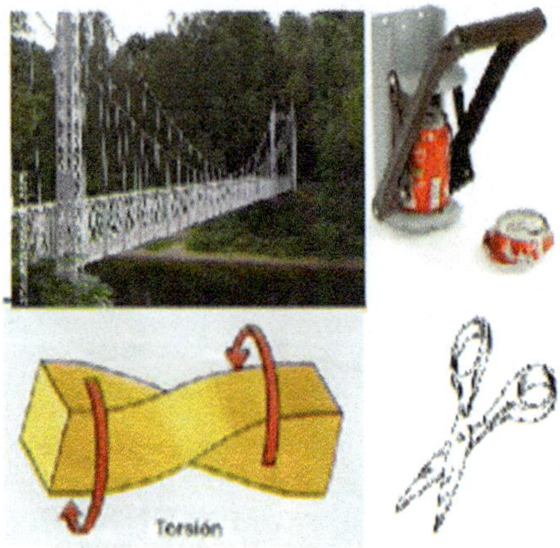

Fig. 6.1 Torion box what where why? AskPhysics.com, Learnersdictionary.com, Ebay.com, www.aboutaberdeen.com, composite of four illustrations

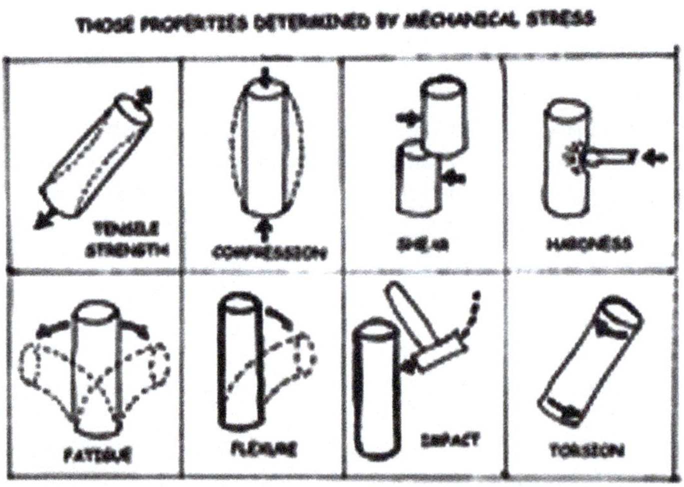

Fig. 6.2 Mechanical properties determined by mechanical stress, *Everything is Material.* Cleveland Ohio: ASM Materials Education Foundation, page 108

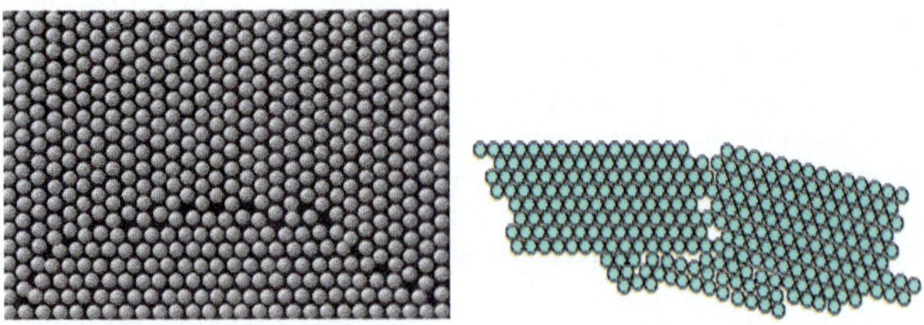

Fig. 6.3 Crystal grains/grain boundaries

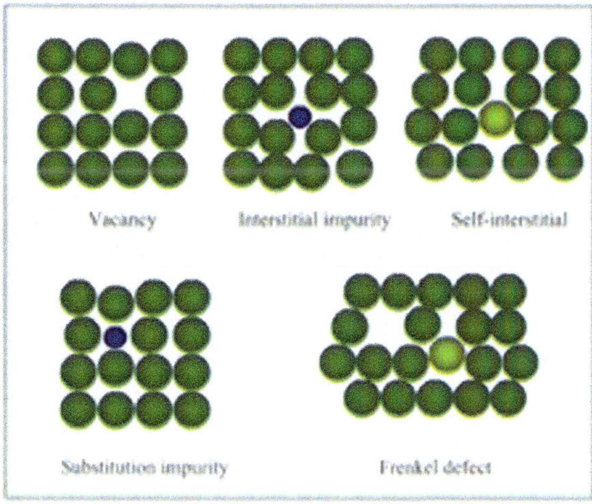

Fig. 6.4 Point defects of crystals

Any piece of metal is made up of a large number of "crystal grains," which are regions of regularity. At the grain boundaries, atoms have become misaligned.

A crystal structure is the alignment of atoms that defines the material. It is a high-ordered structure enabling the structure to be unique. Yet, the crystal structure itself may have defects (Fig. 6.4).

One type of defect, which may occur in a crystalline structure, is called a vacancy. This defect is the absence of an atom in the crystal structure. An interstitial impurity is when an impurity is placed in a space in the crystal structure. If the impurity is placed in an atom's void, the defect is referred to as substitution impurity. A self-interstitial crystal is the placement of an atom within the atomic crystal structure, altering the orientation of the crystal model.

A crystal line defect is the process of forcing atoms together through the process of work hardening. Work hardening may have a beneficial component. When a hot metal is cooled, the metal becomes harder. This is accomplished by strengthening the metal through plastic deformation. This plastic deformation causes a dislocation movement within the crystalline structure of the material. When heat is applied to the arrangement of an atomic structure, the metal becomes softer.

A crystal interfacial defect occurs whenever the crystalline structure of a material is not continuous. A grain boundary must be understood to totally understand the interfacial defect (Fig. 6.5). A grain boundary is the interface between two grains or crystallites. Reference the picture above, the blue lines. The grain boundary is a defect in the crystal structure. It lessens the electrical and/or thermal conductivity of the material (Fig. 6.6).

Work hardening can be experienced in a simple experiment involving steel in the form of "bobby pins" and "paper clips." Label two columns on a sheet of paper, one column listed as "bobby pin" and the other column labeled as "paper clip." Create four rows, indicating one row as a test sample, one row listed as annealed, the third

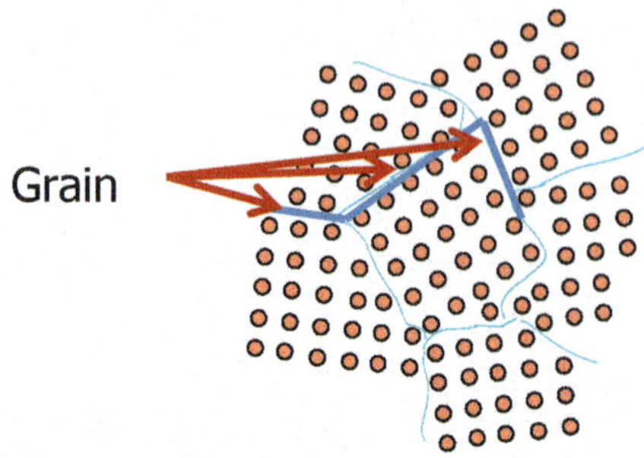

Fig. 6.5 Grain boundaries

Fig. 6.6 Thermal conductivity. Wikimedia Commons, the free media repository (paperclip), en.wikipedia.org (free encyclopedia), Butaneblowtorch.com, composite of three illustrations

row as quenched, and the final row as tempered. Place one "bobby pin" and one "paperclip" as a "test samples" on the paper at the indicated marked area. Using a pair of pliers, hold a "bobby pin" in such a manner that the curved part is placed into a flame. A torch or Bunsen burner may be used as the source of heat. Apply heat until the "bobby pin" is red hot. Place the "bobby pin" on the paper next at the annealed test row. Allow the hot "bobby pin" to cool gradually. This process of cooling is called annealing. Repeat this process with the "paper clip." Place the heated "paper clip" that is being annealed next to the "paper clip" test sample. Taking a third "bobby pin," once again heat until it is red hot. Using the pliers, place this heated "bobby pin" in cold water causing a rapid cool down. This process is known as quenching. Repeat the heating process with a "paper clip." Place the completed samples at the appropriate place on the paper. Be certain to place the tested samples one to the proper row on the paper. The final test is to heat a "bobby pin" until red hot and then quench it. Remove the cooled "bobby pin" with the pair of pliers. Placed the quenched "bobby pin" once more into the blue flame of the torch, but this time take care NOT to allow the "bobby pin" to get red hot (A candle may be used on this portion of the procedure). This new process of heating is called tempering.

It is now time to test the samples. Take the test samples, hold both ends of the "bobby pin," and open the "bobby pin." Describe the resistance, the ease, or the difficulty in spreading the "bobby pin" opened. Record your results. Repeat the same process with the "paper clip." A standard or a reference frame has been created. Repeat the process with the annealed samples, the quenched samples, and the tempered samples, comparing your findings. The purpose of this activity is to observe the effect of various heating techniques. The annealing process makes the material softer by relieving the stress due to the bending of the metal; quenching makes the material stronger, however brittle. Finally, tempering removes the brittleness while maintaining the strength of the item.

A solid material made of organized parts or constituents (i.e., atoms, molecules, or ions) that are microscopically ordered in all directions is called crystals. Many materials are composed of one or more crystals. If the crystal is crystalline, the structure is very orderly and extends over a long range. Crystallography is the study of the arrangement of atoms in solids, the crystal structure of the solid (Fig. 6.7).

This is an important field of materials science due to the fact that crystalline patterns affect the physical properties of materials. A crystal lattice is a three-dimensional arrangement of atoms in a repeating structure (Fig. 6.8).

The formulation of such structures is created by placing one or more atoms at each crystal point. An illustration of a crystal lattice that has two atoms occupying a lattice point is to the right. Notice the two different lattices in the structure (Fig. 6.9).

The smallest piece of a crystal lattice that contains all the information to make the entire lattice is called the unit cell (Fig. 6.10).

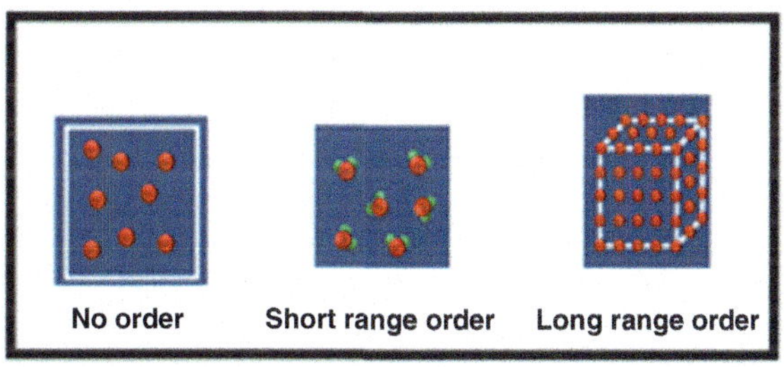

Fig. 6.7 Crystalline materials

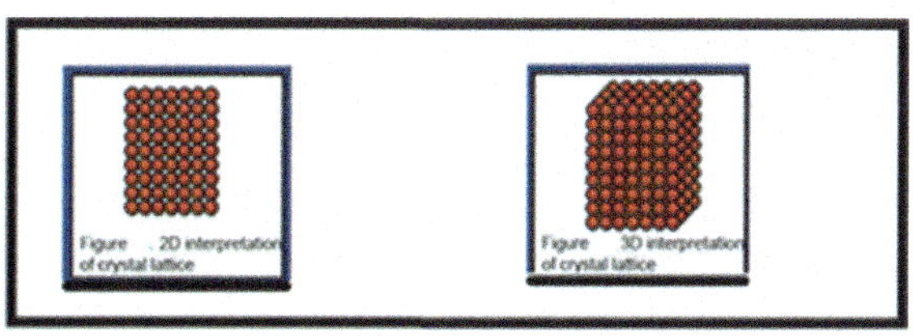

Fig. 6.8 Crystal lattices

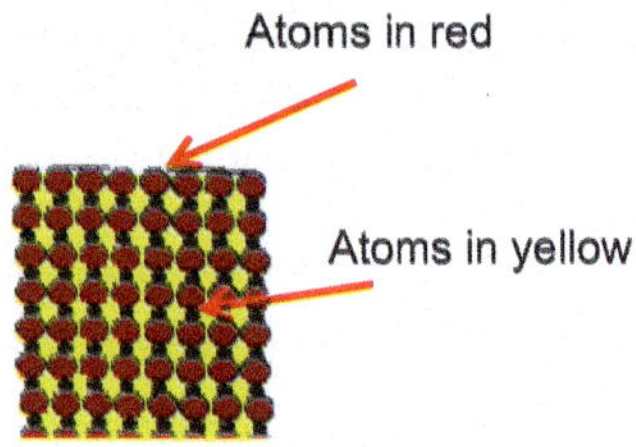

Fig. 6.9 Crystal lattice

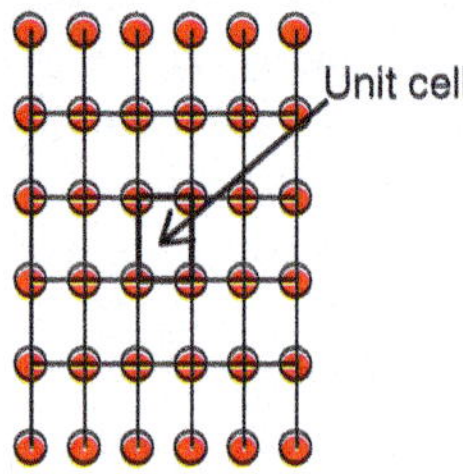

Fig. 6.10 Unit cell

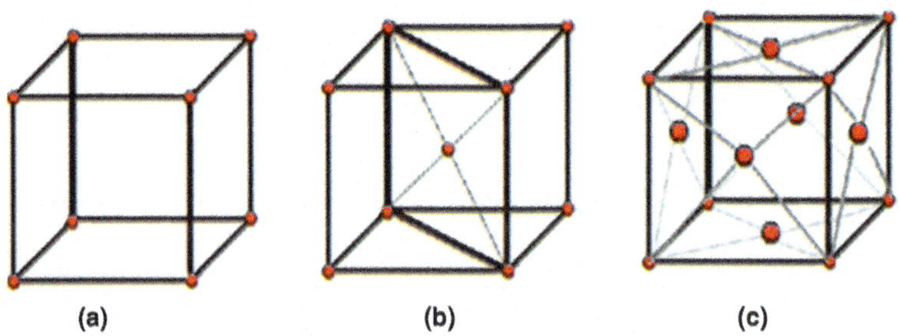

Fig. 6.11 Crystal cubics. (**a**) Simple cubic; (**b**) Body-centered cubic; (**c**) Face-centered-cubic

There are three basic cubic lattices, the simple cubic, the body- centered cubic, and the face-centered cubic (Fig. 6.11).

The simple cubic has a fractional part of an atom at each corner. The illustration (a) depicts an atom on each corner. However, this same structure contains a lattice on the top, on the sides, on the bottom, and in the front and back. In essence, there is one eighth of an atom on each corner. A body-centered cubic, illustration (b), is identical to the simple cubic with the addition of an atom in the center. Placing an atom on the face side creates a face-centered cubic lattice. The face has one-half an atom for each face with the other half being shared by the atom in front, on the side, or the top or bottom. Each lattice has eight corners. The body-centered cubic has one additional atom in the center, and the face-centered cubic has atoms placed on all six sides. There are 14 Bravais lattice structures. The unit cell shape for each crystal system can be described by the relative lengths of the unit vectors and the angles between them. Different lattice forms are possible within each of the crystal systems since the lattice points, within the unit cell, may be arranged in different ways. There are 14 different possibilities in distributing lattice points to construct space lattices. These 14 lattices are referred to as Bravais lattices.

Lattice planes cut through the unit cell in different ways. The Miller indices describe the direction and planes in the lattices.

https://iit.edu/arc/workshops/pdfs/Miller_Indices.pdf

The direction is symbolized by a parenthesis (x,y,z), and the direction is symbolized by the brackets {x,y,z}.

Consider a (110) plane; you would move in the x direction of a unit cell 1, and then move in the y direction of a unit cell and zero in the z direction (Fig. 6.12).

The shaded plane is (110), and the plane is in the z direction. Other important crystallographic plane is the cubic structure. There are many more crystallographic planes (Fig. 6.13).

The fracturing of a metal occurs after plastic deformation has been reached. While applying a stress or a force to a metal by pulling it, the metal elongates. If the metal returns to the initial state after the force ceases, the metal is within the area of elasticity. However, if the force applied stretches the metal beyond the elasticity, changing the metal characteristic, it enters an area of plasticity. Plasticity is a change in the elastic characteristic of the metal. If further stress is applied to the metal, the metal will snap or break. This is known as plastic deformation. A ductile fracture occurs when the metal reaches plastic deformation after a large amount of force or stress has been applied. If a small amount of force causes the plastic deformation,

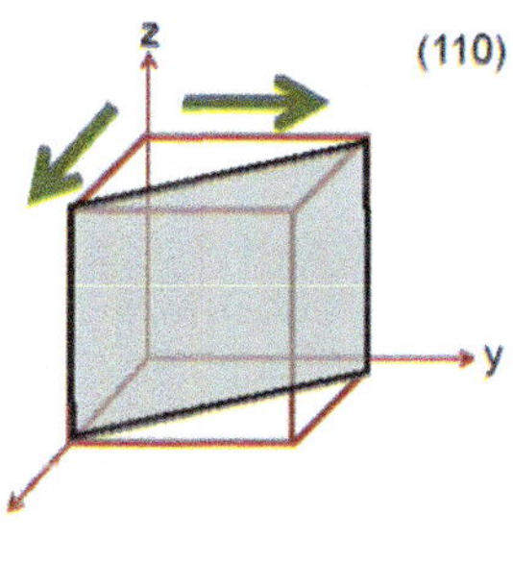

Fig. 6.12 Bravais lattice

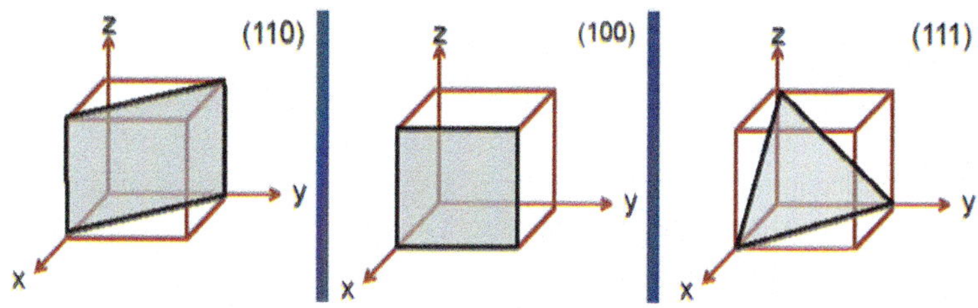

Fig. 6.13 Three Bravais lattices

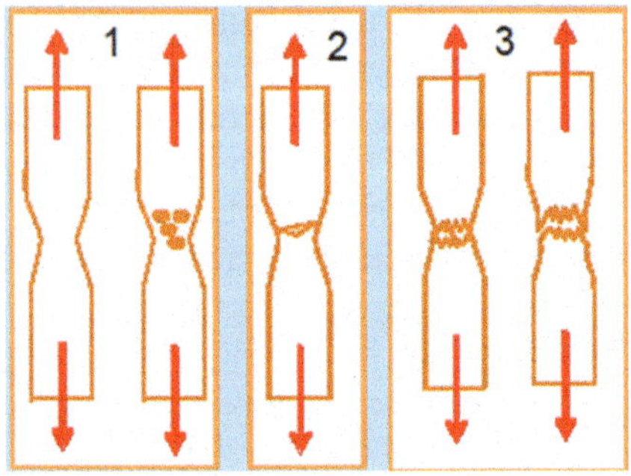

Fig. 6.14 Dogbones, illustrating the necking results during a tensile test

the metal is said to be brittle. During the process of applying a force, the metal goes through a *necking process*. The *necking process* is when the metal thins in the center of the metal and then works its way out uniformly across the metal in both direction and then repeats the thinning through the middle (Fig. 6.14).

In a ductile fracture, three distinct phases occur:

1. The sample forms a neck, and cavities begin to form in the neck.
2. The cavities in the neck coalesce into a crack and near the surface.
3. The crack reaches the surface, and the crack changes to 45° to the tensile axis.

The brittle fracture usually proceeds along the crystallographic planes of the material (refers to a cleavage plane) often as a result of a defect in materials. The brittle fracture also has three distinct phases:

1. The plastic deformation concentrates dislocations along the slip planes.
2. Shear stresses build up nucleating micro-cracks.
3. Further stresses build into micro-cracks.

Creep is a form of a fracture that occurs over time. When a metal experiences a constant load over time, the metal would undergo plastic deformation. The exposure is a function of time, stress, and temperature. The reason why creep is important is that understanding this limitation assists in the selection of the proper material needed for the task in question. The major difference between a fracture and creep is that a fracture can occur suddenly, and creep occurs over a period of time.

Temperature has a major effect on creep. Deformation in a material is noticeable when 30% of the melting temperature is reached. The defect or creep is increased as the temperature increases. Plastics and low-melting temperature metals such as solder creep at low temperature.

A creep demonstration with supporting materials is found in Appendix C.

The jet engine industry is very concerned about creep because of the high temperature that the engines operate. Similarly steam pipes would be subjected to creep. Finally consider a simple incandescent light bulb. As the bulb radiates with heat causing the glow, the filament sags because of creep which causes the light to fail.

Forging is the manufacturing process of applying localized compression forces to form materials. Temperature may be applied to enable the forging, but not all materials demand the heat. Forging has heat applied but does not melt the metal or have the metal cast. One reason to forge a material is that the material made is stronger than any material made another way. When reliability and safety are of concern, forging is the process used. Forging dates back to the blacksmith and has evolved to a highly technical manufacturing process, yet maintains the same concepts of hammering and molding. There are approximately 45,000 people employed in the forging industry across the USA, Mexico, and Canada. It is a $6 billion dollar industry in yearly sales.

Chapter 7
Semiconductors

Living in an electronic society, the cell phone, the computer, and the calculator are in common use. One of the important parts of these electronic gadgets is a semiconductor. Studies in basic electricity investigate conductors and insulators, but what about the substance that falls between these materials, such as the semiconductors? What makes these materials so unique; does the material need to be altered to make it functional? Let's explore!

Fig. 7.1 Conductor, www.Clipartpanda.com

In the study of electricity, the terms *conductors* and *insulators* were introduced (Fig. 7.1). A *conductor* is a material through which an electrical charge may flow from atom to atom with an application of a voltage. An *insulator* functions in the exact opposite manner. An *insulator* is a material which impedes or lessens the flow of electrons. An *insulator* may permit flow of electrons; however, the flow will be weak/poor. When we refer to a *semiconductor*, we are considering a material which exists between the conductor and the insulator. *Semiconductors* have less conductivity than metals. A *semiconductor*, when heated, increases its conductivity, the opposite of a metal when it is heated. Fewer atoms exist in a semiconductor, resulting in less electron movement. **A result of this lessening of electronic movement is known as** ***ionization***. Ionization is the process whereby a *semiconductor* gains or loses electrons from an atom or molecule forming ions, electrically producing an unbalanced element(s) or molecule(s).

A *semiconductor* is not an effective conductor as is gold or copper. However, when the temperature increases to room temperature or higher, more electron

R. A. Wesolowski et al., *The World of Materials*,
https://doi.org/10.1007/978-3-030-17847-5_7

mobility occurs. This implies that electrons travel at a greater (speed) than they do through metals at the same higher temperature. Semiconductive devices can display a range of useful properties. One such property is the directional movement of electrons, i.e., electrical current can pass more easily in one direction than the other. Semiconductive materials disclose a variable resistance and sensitivity to both light and heat. Semiconductive materials can be modified or amended by adding impurities or by being affected by electric fields. Thus, a semiconductor may (can) be used for amplification and directional switching.

Adding impurities to the semiconductor is referred to as *doping*. This process of doping a semiconductor increases the charged particles within the material. When an impurity is added to a pure semiconductor, it is referred to as an intrinsic semiconductor. When this occurs, the pure semiconductor's characteristics are altered. The addition of impurities may increase an electron's velocity by modifying the frequency of the semiconductor. This addition of impurities to a semiconductor allows the electrons of the semiconductor to travel in an environ, creating a high-frequency electrical field. The impurities used in this doping process are dependent on the type of semiconductor. As the impurity is added to the semiconductor, the material may have an increase in electrons. When an increase of electrons occurs through the process of doping, an *n-type* material is created. When a semiconductor is doped heavily, creating a behavior similar to that of a conductor rather than a semiconductor, the process of doping is referred to as *degenerate*.

Extrinsic refers to semiconductors that are lightly and/or moderately doped. As the impurity is added, this composite material may be deficient in the number of electrons and accept electrons from its host material. The result is a positively charged material, deficient of electrons and appearing to have a void where (the) electrons may have been located. This void or hole forms what is called a *p-type* material. In physics, a hole is a positively charged electric carrier. It has the same magnitude of an electron, but it has the opposite polarity. Holes should not be misunderstood (to be) as positrons, which is a particle in antimatter.

Both holes and electrons are located in a semiconductor. The electron flows from negative to positive, and the holes flow from positive to negative. We can better understand the creation of a semiconductor through the doping process. Recall the semiconductor falls between the ability of a conductor which allows electrons to flow readily and an insulator that is a poor conductor. (The) A semiconductor has an atomic structure that allows the conductivity to be controlled through electric current, electromagnetic fields, and even light. This allows the semiconductor to be used in converting light into electricity, as well as electricity into light.

Silicon, carbon, and germanium possess a unique property. They each have four electrons located within their valence shell. This allows the elements to form a perfect covalent bonds, forming a lattice with the neighboring elements. Carbon, in an organized form, crystalline in structure, forms a diamond. Silicon, in its crystalline forms, creates a silvery metallic material.

These free electrons are capable of flowing around atoms, and this movement of electrons results in electricity. Metals which have this property make excellent conductors. Silicon is just the opposite. It behaves more like an insulator retarding the flow of electrons. Silicon allows very few electrons to flow through it. This can be modified by adding impurities to the silicon, previously referred to as **doping**. As previously mentioned, there are two types of impurities, the *n-type* and the *p-type* impurity. The *n-type* impurity creates an excessive amount of electrons. When doping is performed, adding arsenic or phosphorous to the silicon, there is a sharing of four electrons between the silicon and the doped material, allowing the excessive electrons to flow freely through an atom. This electron movement causes an electrical current to pass through the silicon with a negative charge, an *n-type*.

If boron was used as the doping material, there would only be three electrons in the outer shell. When bonded to (the) silicon, the lattice structure which would be formed would be lacking an electron, creating a hole to exists where the silicon electron has nothing with which to bond. With this void of an electron, the behavior is such that it appears to be positive, hence *p-type*. The hole can accept an extra electron from a neighboring atom resulting in a movement of the hole. This makes the *p-type* an excellent conductor.

A small amount of doping material, *n-type* or *p-type*, makes silicon, which is a near insulator, into a conductor, although weak yet nevertheless a conductor. It is significantly better than before the doping, now a semiconductor. The *n-type* and the *p-type* semiconductors by themselves are not that significant. However when they are joined to form a **P-N junction**, they become significant at the junction of the semiconductor.

What makes the **P-N junction** unique is that the material does not allow for the electrons to flow in either direction. It restricts the flow in one direction forming a basic diode. When the electrons flow from the *n-type* region across the *p-type* region, the electrons combine with the holes. Reversing the direction of the electron flow, the electrons and the holes are forced apart preventing electron flow across the junction. The electron flow of the two events is opposite to each other or 180 degrees to each other. If the electron flow is in some other direction than the two directions mentioned, it will allow for the creation of other components such as transistors. A **transistor** is created by having three **P-N junctions**. When using only two *p-type* or *n-type* semiconductors, a simple **P-N junction** is formed by combining two *n-type* and one *p-type* semiconductors or one *n-type* and two *p-type* semiconductors. A **P-N-P** or a **N-P-N** transistor is created, known as a **bipolar junction transistor (BJT)**. The **BJT** can block a current flow, or it can enhance the current flow acting as a switch.

When an impurity is exchanged in the semiconductor, the entire characteristic of the semiconductor is altered. The change may increase the amount of electron flow, or it may decrease the quantity of electron flow. An extrinsic semiconductor is a material wherein the electron flow is increased. The impurity is introduced to the crystal lattice, and, in doing so, the electrical properties are changed, increasing the electron flow. The mobility is also dependent upon the temperature. The lattice scattering is a result of high temperatures. At high temperatures, there is the thermal vibration of the atoms, causing an obstruction in the electron movement. An ion scattering occurs at low temperatures and is dependent of the doped ions. The ions, which behave as scattering centers, prevent the electron movement toward higher energy levels due to the attractive force of the ion and its carrier. This results in an attraction as seen by Coulomb's force:

$$F = \left[k q_1 q_2\right] / d^2$$

In this equation, the *q's* represent the ion charge and the charge on the carrier. The *d* is the distance of separation, and the *k* represents the permittivity constant.

All materials that have electrical properties can be divided into one of three distinct categories: conductors, insulators, and semiconductors. Metals, pure metals, and their alloys are the most part conductors of electricity, readily allowing electron flow. Nonmetal elements and their compounds fall in the category of insulators. They are poor conductors of electricity. The final group of materials called semiconductors fall into a group which are between conductors and insulators. Although elements such as silicon fall into this category of semiconductors, it is not until the semiconducting material has impurities add to it that the effects of electron flow become significant and the directional flow of the electrons is dependent on what impurity is added. Semiconductors are used extensively in devices such as computers and calculators.

Diodes, transistors, and rectifiers are material parts used in the world of electronics. These electronic parts rely heavily on the semiconductor that creates the material. The world in the twenty-first century relies heavily on this material that isn't a great conductor nor a great insulator. It is merely a material that has transformed our way of living.

Chapter 8
Iron Wire Lab and Crystal Models

When changes occur, oftentimes the change is visible. But there are changes that occur while the physical property remains unaltered. Materials that experience phase changes, transitioning from BCC to FCC, may have experienced a transition from a ferromagnetic state to a paramagnetic state without altering the appearance of the material.

Iron Wire Lab

When changes occur, oftentimes the change is visible. But there are changes that occur while the physical property remains unaltered. Materials that experience phase changes, transitioning from BCC to FCC, may have experienced a transition from a ferromagnetic state to a paramagnetic state without altering the appearance of the material.

Many changes that occur may be noticed or visible: a glass that has been dropped and is broken, the painting of a car as viewed by the owner, or the rearrangement of furniture in a room to the homeowner. There are other changes that are not as visual or noticeable: the increase of memory in a computer, or a hip replacement as seen by a passerby, the alteration or change may not be apparent. Transitioning from one phase state to another may not be noticeable or undetected, but the events occur daily. When discussing a body-centered cubic or a face-centered cubic structure, being aware of what the effect of the cubic structure is on a crystal is significant. Consider the body center cubic (BCC), there is an eighth of an atom on each corner of the cube and a full atom in the center. There is space or unfilled area between the atoms, or 68% of the structure is filled. In the face-centered cubic (FCC), there is an eighth of an atom on each corner of the cube and one half of an atom on each face of the cube. The void is lessened, or 74% of the crystal structure is filled. Knowing this fact is critical. The more densely packed crystal, less void, is stronger than the less dense crystal.

Crystalline structure contributes to the composition of materials. When crystal orientation is closely packed, then there is an easier capability to have planes slide past one another. This implies that a closely packed crystal structure allows for a

R. A. Wesolowski et al., *The World of Materials*,
https://doi.org/10.1007/978-3-030-17847-5_8

greater opportunity for plastic deformation. Additionally, cubic lattices allow for a great slippage than non-cubic lattices allow. The face-centered cubic (FCC) has more susceptibility to ductility (i.e., the ability to change form without breaking (Dictionary.com 2015)) than body-centered cubic (BCC). Body-centered cubic crystals have a greater spacing between the atoms, loosely packed, causing the material to be much stronger.

Body-centered cubic materials are more temperature dependent than face-centered cubic materials even at low temperatures. This is perhaps a result of interstitial impurities that are placed in the spacing in the lattice structure.

A lab experience called the iron lab experiment helps illustrate the transitions from one phase change to another. Consider the following materials are needed for the experiment: variac (device that provides an adjustable voltage source of alternating current), meter stick, ring stands (2), magnet, and a weight (~100 g mass), and assemble the materials in the manner demonstrated below (Fig. 8.1).

Slowly increase the setting on the variac from 0 to 30 (measured in percentage). This will increase the voltage through the wire causing the wire to heat. This heating will excite the molecules in the wire resulting in a slow expansion of its length. What will be noticed is that the wire will begin to sag near the 100 g mass. The amount of sag can be measured using the meter stick. Having measured the original length of wire, using algebra and the Pythagorean theorem, the expanded length can be determined. Further increases in the voltage will render an increase in heat in the wire causing it to radiate a red color with heat. The degrees of heat are measured in color ranges transitioning from blue, red, and yellow to white hot. Blue is the coolest range of heat to white which is extreme heat. As the heat is further increased, it

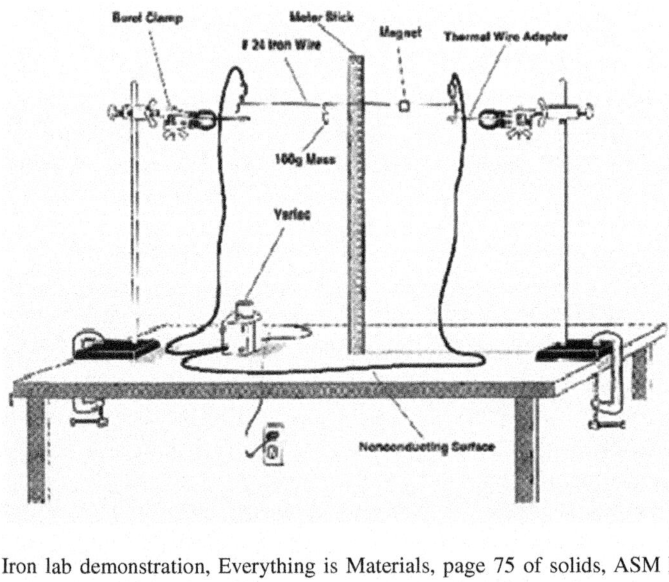

Fig. 8.1 Iron lab demonstration, Everything is Materials, page 75 of solids, ASM Educational Foundation

reaches a point where the magnet falls from the wire. This occurs at the Curie point ~770 °C, depending on the grade of iron. The Curie point occurs when the ferromagnetic state evolves to a paramagnetic state in the iron wire due to added heat (Dictionary.com 2015). The ferrimagnet is an iron magnet which does not need a magnetic field to exist, and a paramagnet is an iron magnet that needs a magnetic field to exist. The magnet fell because there was a transition from the ferromagnetic state where magnetism exists to the paramagnetic state where it does not. As heat is raised, the radiation of the wire becomes bright red. A warning must be made here. If excessive heat is applied to the wire, it will become brittle and may break. One's hope is to turn off the variac prior to this situation occurring. As the wire cools, the wire's length is reduced. Careful observation of the wire will reveal a constant shrinkage of the wire, and then there is a sudden stop to this process. The wire appears to be pulsating for a moment. Afterward, a reduction in the length continues. This fluctuation is a result of transitioning from the FCC state to the BCC state. As heat was applied, the transition was not noticed due to a gradual temperature increase. When cooling ensues, the transition is dynamic and noticeable.

This demonstration provides a source of data validating linear expansion. This provides us with enough information to solve for the coefficient of linear expansion. The process illustrates the Curie point, namely, a phase change from ferromagnetic to paramagnetic, the transition from BCC to FCC, as well as the reverse transition from FCC to BCC.

Addition analysis must be made regarding crystals. By definition, a crystal is a group of atoms that form a repeating pattern. A grain is an individual crystal of many crystals. Being randomly orientated, the interface between two grains in a polycrystalline material, creates a grain boundary.

One type of grain growth is a "dendrite crystal." The growth of a dendrite is similar to branches of a tree, and their growth is in varying direction. The growth will continue until the grain interferes or inhibits the growth of another grain. The space between the crystalized arms allows for the growth of additional regular crystals (Fig. 8.2). A nucleation point is the onset of grain growth. The grain growth may lead to the formation of additional dendritic arms. When complete the dendrite arms join together to create a total crystal structure. The formation of ice on a window is an example of dendritic growth.

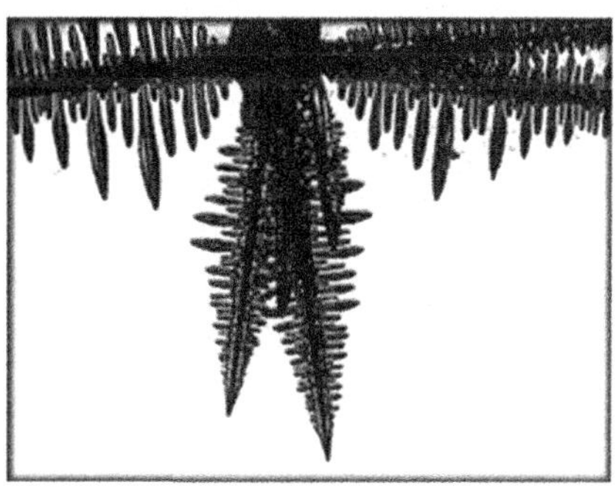

Fig. 8.2 Dendrites, Everything is Materials, ASM Educational Foundation

Fig. 8.3 Sulfur demonstration, CMU Teacher Camp, 2013

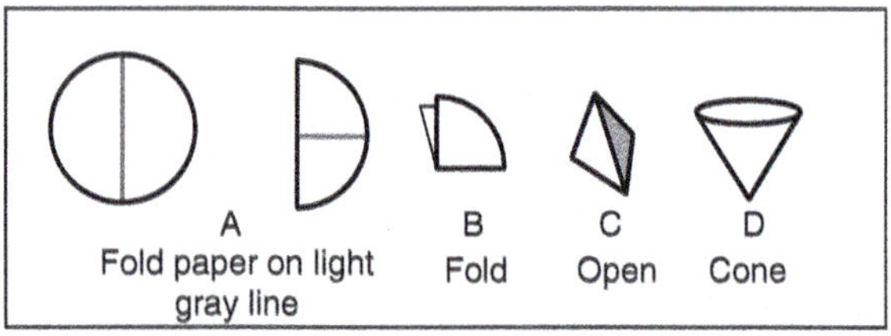

Fig. 8.4 Preparing the filter paper

There are various methods in creating a crystal. Cooling a liquid from a melt is one way to make a crystal. Growth from a precipitate (i.e., solid formed from a liquid (Dictionary.com 2015)) is another method of crystal formation. As a solvent cools from a solution, crystals may be formed. A simple crystal formation created from a solution where the solvent cools or vaporizes.

Place sulfur into a test tube approximately one third full and heat the test tube with a torch or Bunsen burner slowly, rotating the heat to prevent burning (Fig. 8.3).

The heating is melting the sulfur while maintaining a yellow consistency. Place the liquid onto folded filter paper, the fold is in quarters; making a funnel, this paper is placed in a funnel and one quartered opened (Fig. 8.4).

The hot sulfur is poured onto the filer paper. Once a crust begins to form, open the folded filter pater and allow the sulfur to cool (Fig. 8.5).

The crystal structure can now be observed.

Sulfur behaves chemically in one of two ways. A sulfur atom can acquire two electrons to complete an octet in the valence shell. This would form a stable S^{-2}. It can also have a positive oxidation state especially when it is placed in an environment of oxidizers of fluorine or oxygen. Sulfur readily burns in the air forming a sulfur dioxide gas. This formation is a result of covalent bonding between the sulfur and oxygen.

Fig. 8.5 Sulfur crystal, CMU Teacher Camp, 2013

Fig. 8.6 Sulfur crystal monoclinic, CMU Teacher Camp, 2013

If sulfur reacts to metals (or hydrogen), it becomes the more electronegative partner in the bonding and thus becomes a −2. Whereas, if sulfur reacts to fluorine or oxygen, its oxidation number can be +1, +2, …, +6. The most common of these is +4 occurring in SO_2 and SF_4.

The crystal formed by heating the sulfur and pouring onto a filter paper is monoclinic sulfur. It is accomplished by allowing the sulfur to cool and crystallize. The sulfur crystal is stable above 96 °C and becomes unstable below this temperature reverting to a rhombic form (Fig. 8.6).

The second variation of crystal growth is through a chemical reaction process. If silver nitrate is placed near copper, a formation of a silver nitrate and copper nitrate is produced. This crystal form is a dendrite (Fig. 8.7).

A third kind of crystal that may be produced is known as amorphous sulfur. Placing sulfur in a test tube approximately one third full, heat is applied with a propane torch or a Bunsen burner. The sulfur begins to liquefy. Heat is maintained until the sulfur changes state, moving from the solid to a liquid, yet remaining a yellow color, finally transitioning from a yellow liquid to a red liquid stage. The red liquid will begin to change to a maple syrup consistency. Continuing the heat, this syrup will transform into a dark runny liquid. The liquid is then quenched (i.e., rapid cooling) by pouring it into a cool beaker of water. A product with the consistency of a long rubbery, amber-colored strand is formed. Within 24 hours, the amorphous material will begin to return to its original crystalline state. A solid state phase change is occurring.

Fig. 8.7 Dendrite crystal, en.wikipedia.org 1000 × 676 Search by image Manganese dendrites on a limestone bedding plane

The experiments with sulfur lead to a definition. What is an allotrope? An allotrope is "one of two or more existing forms of an element: Graphite and diamond are examples of allotropes of carbon" (Dictionary.com 2015). An allotrope can also be referred to as a polymorphism. It is shaped through a different arrangement of atoms. Examples of allotropes would be carbon as graphite and diamond, oxygen being of the form of O_2 in the atmosphere and O_3 in the ozone, and iron transitioning from BCC to FCC with the inclusion of heat. A change from an amorphous state to a crystalline state may also occur due to a change by heat as has been illustrated in the sulfur experiments.

Reference

Dictionary.com (n.d.). Retrieved March 3, 2015, from http://dictionary.reference.com/

Chapter 9
Graphs and Diagrams

Pictorial representation can help explain the happening of an event. One such representation is a phase diagram. Initially in chemistry phase diagrams are investigated to explain the transitioning from solid to liquid to gas of water. The transition does not occur linearly because of the phase changes that occur moving from the solid state to the liquid state and then from the liquid state to the gaseous state. A graph may help explain this state change. Graphs may also be used to investigate the relationship between terms, force changes to length changes. The slope analysis may help explain Young's modulus. Graphics are beneficial in understanding the happenings in the environment of materials.

Graphing and Diagrams

"Graphs and diagrams can, and usually do, serve two different functions. They can be useful analytical tools for professional statisticians, and they can be ways of displaying data to the public at large. The former include Normal probability plots, boxplots (box-and-whisker diagrams) and correlograms, and it can be expected that statisticians will be able to use them when appropriate and understand them without difficulty. However, it is also an important part of the work of statisticians that we should be able to present data in a coherent and meaningful manner to non-statisticians, and there are a great many methods of doing so. One can hardly find an issue of a serious newspaper that does not include several graphs. A single such picture conveys – or can convey – the important features of the data more vividly and memorably than columns of data." (Graphs and diagrams – Royal Statistical Society, https://www.rss.org.uk/graphs-diagrams).

A pictorial representation of numbers or relationships is found on a graph. Often times a student states they are a visual learner. The combining of these two ideas enables a better understanding of what is occurring. The graphic illustration can depict the highs and lows, the slope of a graph can explain the rate of decline or increase, and any abnormality or abrupt change can become apparent with a graphic

R. A. Wesolowski et al., *The World of Materials*,
https://doi.org/10.1007/978-3-030-17847-5_9

interpretation of numbers. Consider a solid piece of ice below freezing. The temperature of this material is below zero degrees Celsius. If heat is applied, what happens? Often times, students will say that the ice melts. This is incorrect, because the heat merely increases the temperature of the ice, but the substance remains a solid. It is only when the solid reaches zero degrees will additional heat allow the solid to transition into a liquid. However, the temperature does not change until the entire solid becomes liquid. At this time the liquid is at zero degrees, and if additional heat is applied would the liquid increase its temperature reading. The transition from the solid to the liquid state occurs at the fusion state where there is ~80 calories of heat needed to convert 1 gram of a solid to the liquid state for ice to water. This 80 calories/gram is called the Latent Heat of Fusion. Heat is needed to help change from a solid to a liquid, but there is no change in temperature. This is an interesting issue; if there exists a temperature change, then there is an increase or decrease of heat; however, when heat is applied or removed, there may not be a temperature change as in the case of fusion, the transitioning from the solid to the liquid state. So think about a definition of temperature!

While in the liquid state heat is applied, the liquid becomes hotter until it reaches a temperature of 100 degrees Celsius. At this temperature, as heat is applied, the material transitions from the liquid state to the gaseous state without further temperature change until there exists a total state change from liquid to gas. As in the fusion state, an amount of heat must be added, in particular ~ 540 calories of heat per gram of water, to allow this change of state to occur. Consider the severity of a steam burn as compared to a burn from boiling water. The steam has ~540 calories of more heat per gram than a gram of water.

Once the transition of the water into steam has occurred, and additional amount of heat would cause the temperature to increase.

It was previously mentioned that many students claim they are visual learners. Consider the description of the heating of ice until it becomes heated steam as visualized in a graph; it would look like this (Fig. 9.1):

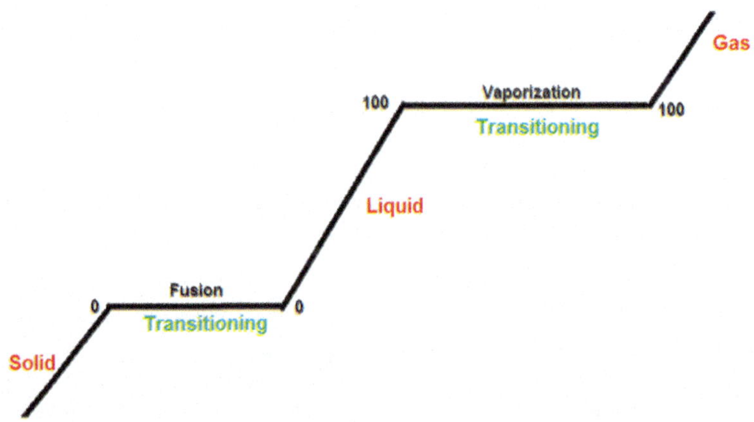

Fig. 9.1 Heating of water from a solid state to a gaseous state. (Self-drawn, Wesolowski 2015)

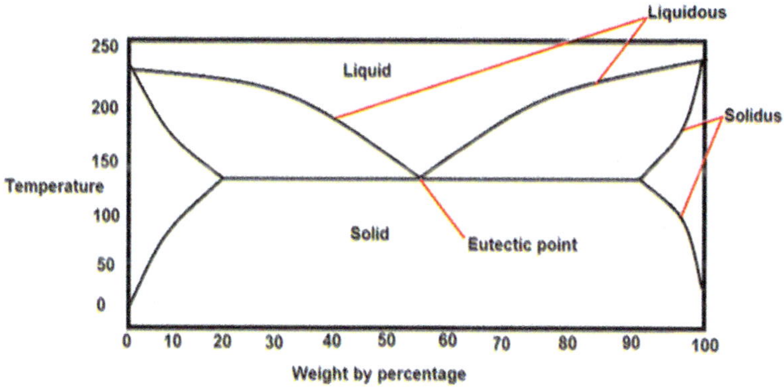

Fig. 9.2 Phase diagram of bismuth-tin mixture. (Self-drawn, Wesolowski 2015)

Would there be an understanding without and explanation? Would the explanation be made clearer with the graphic interpretation? The ice to water to steam transformation is simply in the understanding of the transition when it can be visualized and leads to better comprehension.

Consider a phase diagram of a bismuth-tin mixture (Fig. 9.2).

This particular phase diagram appears to be more complex because the metal is not pure, and the composition by percentage of tin of bismuth changes from 0% to 100%. There are states where the solution is solid and when the solution is liquid. Is the tin always solid when the bismuth is solid? Can tin be liquid and the bismuth be solid at the same time? There are transition points where the material changes states, but the transition may be of one material and not the other in this combination of tin and bismuth. Does the percentage of one material when mixed with another material affect the transition state? There is a point where the material is in a "solidus" and starts transitioning and melting, but it has not melted completely. A similar transition occurs at the "liquids" state. The "liquidous" state is the location where the substance transitions fully to a liquid. Notice on the graph the solidus and liquidous states are not merely a point location but are extended over the distance on a line. This is due to the changing concentration of the percentage of one material to another. The eutectic point is the point of lowest solidification of any material that transitioned from the liquid state. Notice from the graph this occurs only at one point and not over an area or distance.

Graphical interpretation of data or events helps in the understanding and clarification of theories presented. Some individuals may be visual learners, and others may not be; however, the inclusion of a graphic to help visualize the intent of the data enables a better understanding of what is occurring.

Using a supply of tin and bismuth, place 11 cups marked from 0% bismuth and 100% tin to the 11th cup which would read 100% bismuth and 0% tin; the percentage would be determined by mass. If the maximum mass desired is 50 grams, the supply purchased would last for some time. Once the mass is mixed, pour a sample into a predetermined mold so the results poured would be similar to each other, and place them in a marked container. Once all of the samples are made, place the samples on a griddle. If a griddle is not available, use a hot plate, placing the samples along the plate's edge. As heat is applied, the samples will begin to melt. The melting should match the phase diagram of the tin-bismuth chart. This demonstration will reinforce the idea of the phase diagram.

The first sample that should melt should be the eutectic sample, ~50% mixture of each. Following the transitioning from the solid to liquid state based on the phase diagram, observe the order of melting, and predict which will be the last to melt? This is not a perfect lap, the findings may be off, and this may be dependent on the measurements and percentages of the masses. Also, are the samples used pure bismuth and tin? If the material used is an alloy product, the results will be skewed.

Chapter 10
Corrosion

Corrosion is viewed as a negative, something that must be someone's fault. Seldom is this corrosion viewed as a natural process. It appears that all materials wish to return to its original state. When viewed this way, materials that are deteriorating are merely returning to the original state in earth as minerals and ores. Attempts are made to slow the rate of corrosion; the ultimate result will eventually occur.

Corrosion
The deterioration of a material as a result of an interaction between the environment and the material defines a process known as corrosion.

The material that corrodes and the environment, both, affect the reaction time of the corroding process.

The corrosive process that is most familiar creates a red by-product over the original material; it is best known as "rust." This red material is iron oxide (Fe_2O_3) brought about through the interaction of oxygen and iron.

Corrosion can be kept in control, slowed, or even stopped by using proper techniques. Such techniques to be considered to retard the spread of corrosion are:

1. Environmental modifications
2. Metal selection and surface conditions
3. Cathodic protection
4. Corrosion inhibitors
5. Coating
6. Plating

Considering the amount of money, the federal government spends to offset major disasters, hurricanes, tornados, tropical storms, floods, and fires; the approximate cost would be \$17 billion dollars annually. If governmental expense was in the area of \$276 billion dollars, or approximately 16 times the amount of money the government spends on natural disasters, the expenditure would be viewed as monumental. This amount is approximately 3.1% of the gross national product. To what is this expenditure dedicated? The answer is corrosion (Fig. 10.1).

R. A. Wesolowski et al., *The World of Materials*,
https://doi.org/10.1007/978-3-030-17847-5_10

Fig. 10.1 The U.S. military storms to the rescue in foreign disaster relief, govbooktalk.gpo.gov 450 × 291 Search by image, Image: Natural disaster word collage Source: JimKimmartin.com

Fig. 10.2 Invest with Infrastructure, Public Roads, Vol 68, No 2 September/October 2004 https://www.fhwa.dot.gov/publications/publicroads/04sep/08.cfm (Boscarino, T., 2013)

Corrosion is a natural phenomenon. It is an electrochemical reaction between a metal and its environment. Corrosion is the tendency of a metal's return to its original state. Corrosion involves the flow of direct current (DC) (Fig. 10.2).

A great deal of money is dedicated to this natural occurrence called corrosion. The federal government speaks of the deterioration of the infrastructure of our society. This can be noted in the 2013 Report Card for American Infrastructure by American Society of Civil Engineers (ASCE), awarding the infrastructure a D+, a follow-up report by the same organization rating the bridges in America a C+. The Senate Budget Committee, headed by ranking member, Bernie Sanders reported "Unfortunately, whereas once the United States could easily boast of having the most sophisticated, modern infrastructure in the world, today we are rapidly falling behind. We simply have not been making the kinds of investments that are required to even maintain the system we have now, let alone improve and update it. Instead of maintaining and investing in a more efficient and effective system, our infrastructure has deteriorated because of insufficient funding and misplaced priorities" (Senate Budget Committee Bernie Sanders, Ranking Member (n.d.)). There exists a need to rebuild this infrastructure. Ironically 25 percent of the expected cost to repair the deterioration could be saved using existing technology (May, n.d.). Given this information, how concerned is the industrial community or the government in this issue of corrosion? Corrosion is found in pipelines, underground storage tanks, industrial plants, power generation, and transportation. Corrosion is found at marine areas, waterworks, petrochemical plants, and pulp and paper mills, in real estate and in private homes. Corrosion exists everywhere. Corrosion appears to be a negative process. But could it be essentially good?

A battery consists of an anode, cathode, electrolytic material, and metal. An anode is the part of the battery where oxidation (i.e., reaction in which the atoms in an element lose electrons and the valence of the element is correspondingly increased (Dictionary.com)) occurs. Electrons are released. There is a tendency for metals to lose electrons to become more stable, removing excessive electrons from its outer shell. A cathode is the portion of the battery where reduction (i.e., a chemical change where electrons are gained (Myers et al. 2006)) occurs. Electrons are received. The metal becomes less stable establishing a higher energy level. The metal is the avenue through which electron transfer takes place, the movement from the anode to the cathode. The electrolytic is a material that completes the circuit, allowing electrons to flow. The electrolytic may be salt water, an acid, or a paste (Fig. 10.3).

Having described the components of a battery, how does this battery work? There is a chemical buildup of electrons in the battery in an area called the anode. There is instability within the battery between the anode and the side that is without an excess of electron. The electrons want to rearrange themselves to eliminate this difference in charge. This is done in a particular manner. The electrons repel each other attempting to find a location with fewer electrons. In a battery the only place to go is the area of the cathode. The electrolytic material prevents the electron from moving directly to the cathode area. When a circuit is complete or connected, a metal wire links the anode to the cathode area allowing an electron transfer to occur. An electrochemical process occurs between the anodes and cathodes hindering the supply of electrons. As a result, the battery is limited.

Understanding the workings of a battery is vital in comprehending the process of corrosion. A "corrosion cell" is an electrochemical reaction involving a transfer of electrons, an oxidation process whereby electrons leave a source through a reduction process with electrons being received. There is a migration of electrons. This

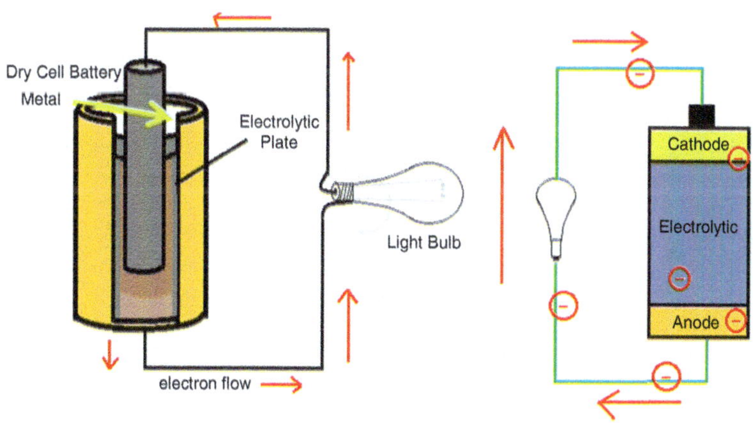

Fig. 10.3 ACME – A Battery. (Self-drawn, Wesolowski, 2015)

process occurs through a material which completes the circuit. Given a metal such as iron, electrons may be transferred from the iron (anode) to another element or compound such as water H_2O or oxygen O_2 (cathode). A metal involved in the corrosion process must be willing to relinquish its electrons. As this process occurs, a new material is being formed. This new material may be a protective material or it may react again (Fig. 10.4).

Corrosion is a form of distress in a material and happens prior to the object's failure. The process of corrosion is one form of distress, a form which can be controlled if properly treated. This deterioration may be caught at an early state when first recognized and treated to prevent the material's failure. Corrosion's prevention is available.

There are various forms of corrosion: a pitting, stress cracking, and uniform corrosion. The causes of corrosion are numerous (Fig. 10.5).

Fig. 10.4 Metallurgy and material engineering. (Self-drawn, Wesolowski 2015)

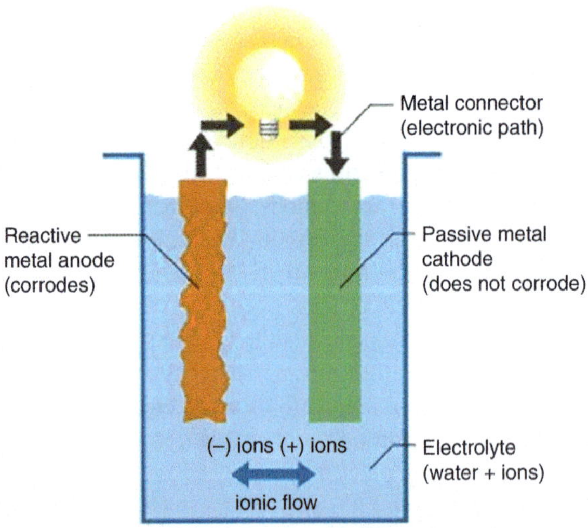

Fig. 10.5 Corrosion types, Pitted image from St Joseph HS corrosion project by students, General Corrosion, www.pelliottcorrosion.com 700 × 525 Search by image, Crevice Corrosion, corrosion.ksc.nasa.gov 684×489 Search by image. (Pictures adapted by Wesolowski 2015)

Use of dissimilar metals is a common source for corrosion. The reaction between the different metals, electron transfer, initiates the deterioration based on the oxidation state of the given metals. Bacteria are a major cause of corrosion in metals. There are dozens of microbiological organisms that have major effects on carbon steel, aluminum alloys, copper alloys, and stainless steel, provided a ph factor (i.e., the value used to determine the acidity or alkalinity of a system (Meyers et al. 2006)) between 4 and 9 and temperatures between 10 °C and 50 °C (WebCorr Corrosion Consulting Services, n.d.) exists. Temperature also affects the rates of corrosion. As the heat rises, the corrosion rates increase.

Attempts are made to control the corrosive process. One process of prevention is coating. An example of this process would be by protecting the material subject to corrosion by applying paint or a lacquer to its surface. Galvanization is applying zinc over iron or steel. Aluminum produces its own coating oxide naturally. Using a sacrificial, a material that has higher oxidation than the material to be protected permits the sacrificial to corrode rather than the metal. The sacrificial is the use of a more reactive metal in place of a less reactive metal. Attempts to control the environment may retard the corrosion process, or discretion in the selection of the material may slow the corrosive process.

References

APME. (n.d.). Report on Polymer Identification Codes for Packaging, www.Plasticseurope.org, Retrieved from Ref: apme identification Report 0114

Dictionary.com (n.d.). Retrieved March 3, 2015, from http://dictionary.reference.com/

Myers, R., Oldham, K., & Holt, R. (2006). *Holt chemistry*. Orlando: Holt, Rinehart and Winston. (Meyer, 2006)

Sass, S. (2011). The substance of civilization materials and human history from the stone age to the age of silicon. New York: Arcade Pub.

Senate Budget Committee Bernie Sanders, Ranking Member. (n.d.). Retrieved February 12, 2015, from http://www.budget.senate.gov/democratic/public/index.cfm/repairing-our-infrastructure

Chapter 11
Introduction to Ceramics

Ceramics is usually understood to mean pottery and is associated with jars, cups and plates. Dating back to 24000 BC, ceramics has a history that is interwoven with mankind. There has been a steady increase in the technology and application of ceramics. What is this thing called ceramics, and how is it used will be investigated? Looking at ceramics as a glass, a brief review in optics will be necessary.

Glass and Ceramics

An inorganic, nonmetallic solid material made of metal, nonmetal, and metalloid atoms bonded together by ionic or covalent method. The result is a crystalline structure resulting from a gradual cooling technique. If the cooling technique is rushed creating an amorphous crystal, the result is a clear glass-like material. Many of the glass labs were adapted from ASM International's Teacher Manual and MAST Handbook, http://matse1.matse.illinois.edu/metals/b.html

A ceramic is an inorganic, nonmetallic material. The crystal structure of the material may be crystalline or amorphous. "Ceramics are classified as inorganic and nonmetallic materials that are essential to our daily lifestyle. Ceramic and materials engineers are the people who design the processes in which these products can be made, create new types of ceramic products, and find different uses for ceramic products in everyday life" (Johnson).

The term ceramic comes from the Greek term, keramos, meaning pottery clay or pottery, ("ceramic science"). One of the fundamental uses of this material is for building materials and containers. Man, in the early years, discovered that materials could be made from clay dried and hardened by heating the clay in the sun. "As early as 24,000 BC, animal and human figurines were made from clay and other materials, then fired in kilns partially dug into the ground" (Silnes 2014). Ceramic tiles were discovered in the Mesopotamia and India region dating back to 10,000 BC. This discovery indicates that the process of ceramics was understood. Additionally, the property of ceramics was used in the storage of water and food (Silnes 2014).

R. A. Wesolowski et al., *The World of Materials*,
https://doi.org/10.1007/978-3-030-17847-5_11

Glass was believed to be discovered in Egypt around 8000 BC, when overheating of kilns produced a colored glaze on the pottery. Experts estimate that it was not until 1500 BC that glass was produced independently of ceramics and fashioned into separate items (Silnes 2014). Between 50 AD and 50 BC, optical glass was used in the creation of mirrors, lenses, windows, and glass blowing products. This new use of material was spread by the ancient Romans (Whitehouse 1988).

The first ceramic composite, porcelain, was created by the Chinese around 600 AD. The material used was feldspar and quartz. "Pottery was valued in the Far East as elsewhere for its utility and relative affordability. Early on, however, the Chinese developed a heightened sense of aesthetic appreciation for ceramics that promoted the sensitivity to material, technical sophistication and stylistic range that defined the classic wares of the Neolithic cultures, Bronze Age, T'ang (618–906), Sung, and Ch'in dynasties" ("Guide to Chinese").

During the 1870s and 1880s, new advances in ceramics occurred. In 1870, during the Industrial Revolution, refractory materials (i.e., materials able to sustain extreme temperatures) were introduced. Due to the demands of the Industrial Revolution, industry and society craved ceramic products fashioned from natural raw materials. Artifacts ranged from high-voltage insulators for power lines, wash basins and toilet bowls to ceramic tiles and ceramic sewer pipes, resisting the acidity of the soil (Pampuch 2014).

In 1879, Thomas Edison developed the incandescent light bulb. Many historians agree that the light bulb had been around in 1802, but Edison make the light bulb commercially viable ("Edison muckers"). For more than a century, Americans believed that the incandescent light bulb was the greatest invention to mankind, as indicated in Ernest Freeberg's book, *The Age of Edison: Electric Light and the Invention of Modern America* (Freeberg 2013). The "light bulb" was a symbol of a great idea. In short the light bulb increased the pace of city life and contributed to the expanse of the Industrial Revolution. "By tracing the role of the electric light played in the pivotal decades when our modern culture was born, we can better appreciate that inventions are not simply conjured up by great men like Edison but evolve as they are shaped by a variety of political, economic, and cultural forces" (Freeberg 2013).

In 1889 the American Ceramics Society was founded by R. Butler and S. Adams (1998). Its purpose is to promote the use of ceramics in the industrial and manufacturing business world. During the twentieth century, modern advances in ceramic engineering grew. Fiber optics, photovoltaic cells, superconducting ceramic oxide, and the discovery of "smart" materials were engineered or invented.

The question arises: What properties do ceramic materials possess? Ceramics are poor conductors of electricity and poor conductors of heat. They possess a low density, compared to other materials, and have a high melting point.

Technically, ceramic material is a product that is created when heat permanently changes the material. All clay is a ceramic material, but there are other materials which are not clay that are essentially ceramic. A glaze is a ceramic because the glaze, when heated, is permanently changed.

When heated, the ceramic experience property changes. Heating the material without liquefying the material allows for the elimination of impurities like water. Firing can fuse particles together, allowing pores or voids to be created through the evaporation of water or other materials contained in the clay. The final product becomes brittle.

Why use the ceramic material? Ceramic is harder and stiffer than steel. It is more resistant to corrosion than metals and is less dense than most metals and alloys. The manufacturer relishes the use of this material due to its abundant supply and the low cost of producing it. Traditional use of the product can be found in the production of glass windows, bricks, pottery, china, clay products, and cement. There are a multitude of advances for the use of ceramic materials. In the medical field, ceramics are used in orthopedic joint replacements, prosthesis, dental restoration, and bone implants. The military use ceramic components in the ground, air and naval vehicles, missiles, and sensors. The computer industry uses insulations made of ceramics. Resistors, superconductors, and capacitors have relied on ceramic material to withstand excessive heat generated from electronic components. Consider the field of electronic communication. The introduction of fiber optics has increased the amount of data transported, in addition to the speed of data transportation.

When one element reacts with another element, the purpose of the chemical interaction is to become more stable. Metals will react with nonmetals through ionic bonding to create this stability. An example of this is aluminum reacting with oxygen to form aluminum oxide. Semimetals also react with nonmetals to establish stability through covalent bonding. An example of this reaction would be silicon interacting with oxygen forming quartz, SiO_2. Briefly reviewing, ionic bonding is the transfer of electrons from one element to another, a giver (anion) and a taker (cation). This form of transference occurs between metals and nonmetals. A covalent bond is a sharing of electron in an attempt to fill a valence shell and thereby making the combination more stable. This covalent bonding occurs between semimetals and nonmetals.

The heating of the quartz is one issue in the creation of a ceramic and cooling is the other. If the heated quartz is cooled slowly, i.e., annealed, the crystal structure is organized, and a crystalized atomic structure is formed. However, when the heated material is rapidly cooled, i.e., quenched, the crystal structure is amorphous, not organized, and chaotic. Heating and cooling may cause the atomic structure to vary in its configuration. Due to the realignment of the atoms, the spacing between atoms may change causing a vacancy or change in spacing. Furthermore, heating an additional atom may enter into the lattice array. These events lead to defects. An example of this is a point defect. A point defect refers to an additional atom entering into the atomic structure. This additional atom may replace one of the original atoms of the structure. A substitutional defect is defined as the replacement of an atom by another. The additional atom may not find a void to be placed into but rather may find itself placed in between two existing atoms. This defect is referred to as an interstitial defect. If a void exists and is not replaced by an additional atom, this defect is referred to as a vacancy defect (Fig. 11.1).

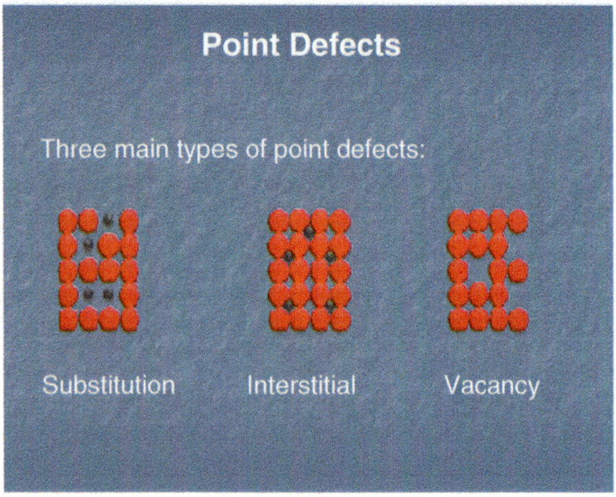

Fig. 11.1 Defects. (Self-drawn, Wesolowski 2015)

A dislocation is another form of defect. A dislocation is described as a position where the atoms are out of position with the crystal structure. Stress is a major cause of a dislocation which results in potential slippage or plastic deformation.

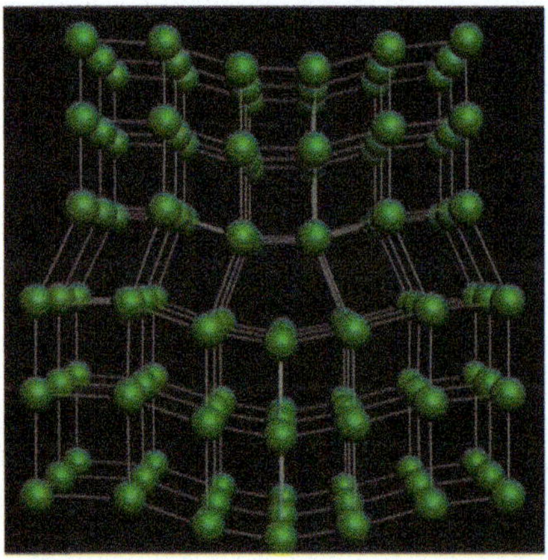

A line dislocation or defect is referred to as the distortion of the atomic structure appears to be linear or on a line, Crystal Defects – Linear Defects (Dislocations) (n.d.).

An interfacial defect refers to the deformation of a planar in nature. The distortion maybe external or internal in nature, pertaining to pores in the structure or on the surface of the material. In the discussion of interfacial defects, grain boundaries become points of interest. Grain boundaries refer to the interface between two grain structures as found in many polycrystalline materials. The grain boundaries have an effect on the electrical or thermal conductivity of the material.

Ceramic material is limited in its use due to its brittleness and microscopic flaws that occur in the cooling process. Unfortunately, these flaws cannot be controlled in the manufacturing process. However, ceramic material possesses vast amounts of compression strength compared to its tensile strength, thereby serving well as construction materials.

Glass is a ceramic product fashioned when heated material is rapidly cooled. This material may be transparent and have optical characteristics. Optical characteristics are important for it allows a method of identifying material. It also identifies certain properties of the material such as color, luster, or brilliance. As light strikes an optical material, light is reflected, diffracted, refracted, or absorbed. Scattering or reflecting light is a surface event. Refraction, diffraction, or absorption of light is internal events. The importance of the optical characteristics of glass can be found in the manufacturing of lenses, mirrors, and "glasses." Fiber optic material is glass material allowing light to travel through the material carrying data files. Light sent through a glass creates a laser beam applicable in medical procedures; storage of energy in solar cells is also a function of ceramic materials.

The wavelength of visible light ranges from 0.40 to 0.75 micrometer (10^{-6}). Within this range the light is absorbed or reflected. It may emit onc color and absorb another. The light beam will travel within the electromagnetic wave, that is, it will have an electrical polarity which is perpendicular to the magnetic polarity. Understanding this phenomena helps explain the working of polarized lenses, where a percentage of the light intensity is reduced. If the polarized lenses are used perpendicular to each other, a total absence of light will occur. Materials are grouped as transparent, translucent, and opaque. The opaque is the absence of all light, the transparent is light being transmitted through an object, and the image of the reflect object is distinct and clear. The translucent material allows light to be transmitted through the medium; however, the reflected image is distorted, or not clearly visible. The light is said to be diffused.

Light that is cast into a material changes its speed because it is entering into a denser medium. If the light were cast at an angle, the path would be bent as a result of its density. This bending of light is referred to as refraction. Refraction is of concern because it may help in defining the property of an object. A jeweler uses an ocular to view the bending light of a diamond. The manner in which the light is bent dictates what the index of refraction is of the material. The index of refraction is a mathematical ratio of the speed of light in a vacuum as it compares to the speed of light in medium being investigated. This refraction of light indicates to the jeweler whether the item of investigation is a diamond or not. Further investigation into this bending of light is accomplished using Snell's Law. This law is the ratio between the speed of light in the air to the speed of light in the medium and the angle of light entering the medium as compared to the angle of light within the medium. The ratio of the comparisons of the speeds or the angles defines the index of refraction of the material.

The application of this bending of a light wave is used in the application of glass and fiber optics, a thin transparent fiber, usually of SiO_2, guiding light from one end of the material to the other. Fiber optics allows for the transmission of telephone, cable, and data to travel at speeds near that of light (186,000 miles/sec). Photovoltaic cells (solar cells) convert light into electricity. Light-emitting diode (LED) allows light to illuminate without a filament. This device lasts longer, is more efficient, and is more durable than a conventional light bulb.

Properties of ceramics are being applied to change the world in which we live. Introduced some 26,000 years ago, their applications are still being altered, from the pottery state to a source of transmission of data. Their resistance to heat has found ceramics as elements of motors in the aeronautics and space industries. Although brittle, the compression force, which this material can sustain, allows its use in a myriad of applications.

References

Butler, R., & Adams, S. (1998). *The American Ceramic Society: 100 years*. United States: The American Ceramic Society.

Crystal Defects – Linear Defects (Dislocations). (n.d.). Retrieved February 12, 2015, from https://www.nde-ed.org/EducationResources/CommunityCollege/Materials/Structure/linear_defects.htm

Freeberg, E. (2013). *The age of Edison electric light and the invention of modern America*. New York: The Penguin Press.

Pampuch, R. (2014). *An introduction to ceramics*. Cham: Springer.

Silnes, A. (2014, May 1). History of Ceramics. Retrieved February 17, 2015, from http://ceramics.org/learn-about-ceramics/history-of-ceramics

Whitehouse, D. (1988, January 1). Corning Museum of Glass. Retrieved February 18, 2015, from http://www.cmog.org/publication/glass-roman-empire-0

Chapter 12
Biomaterials

Why is there a concern on integrating inanimate materials with those materials that are composed of living organisms? Rejection is a major issue when mixing these two types of materials together. Ceramic valves, artificial hips and knees, dental implants, and stints are a few of the materials that are placed in living tissue environments. What types of materials are used?

Biology and ceramics do interact with each other. Ceramic material is used in the body as dental implants and valves in heart surgery because the rejection of the ceramic implant is minimal. In the dental field where the implant replaces the partial plate, a 95% acceptance rate is experienced in the past 40 years (Baylan 2012). More than 90% of patients who have had hip surgery have not needed any additional surgery (Hip Replacement 2013). The need for the additional surgery is not necessarily a direct result of rejection (Jacobson 2013). The artificial hip socket is smaller than the normal hip socket, and as a result the artificial hip may remove itself from the socket. During the beginnings of hip replacement surgery, the implant was designed to be very near the dimensions of the actual hip. It was found that the larger hip resulted in a quicker deterioration as a result of "volumetric wear." "Volumetric wear" refers to the increase of particles exposed on the surface that can wear due to a frictional rubbing. A reduction in the size of the ball reduces this "volumetric wear." To resolve the issue of "volumetric wear," an increase in the size of the ball and socket is further considered with the material of the ball and socket being metallic. However, the metal renders a metallic shaving being introduced into the biological environment, and sensitivity reactions were reintroduced. Material scientist have introduced a plastic or ceramic coating to the ball and socket; however, the results are inconclusive as of this writing (Ganapathi 2010).

Most implant complication fall under four major area: bleeding and surgical complications, postoperative infection, immune response or rejection, and device failure. In all surgical procedures, blood clots may develop during or after surgery. This may increase the possibilities of pulmonary embolism or stroke. Infection is a

R. A. Wesolowski et al., *The World of Materials*,
https://doi.org/10.1007/978-3-030-17847-5_12

possibility after any surgical procedure. A rejection to a graft, implant, or device may occur if there is sensitivity to the foreign material.

> Approximately 13% of the population is sensitive to metals such as nickel, cobalt, and chromium. However, research verifies that less than 0.1% of rejection is due to this metal sensitivity. With an excess of 200,000 hip replacement surgeries performed yearly, 45,000 hip revisions, 400,000 primary total knee replacement performed, and 36,000 knee revisions, the percentage of rejection or complications has been held to a minimum, or the procedure would not be tolerated (Jacobson 2013).

The material selected for the implants must be nontoxic, bio-inert, bioactive, and biodegradable. An implant being nontoxic implies it is safe to the body. It is not poisonous; rather it is harmless, and safe (Dictionary.com 2015). The creator of an implant must be certain that there are no side effects based on the interaction of the material used for the implant and the human body.

Bio-inert refers to the inability to be interactive with biological systems. The bio-inert implant must not respond to the biological environment in which it is placed. It should not have a reaction or be affected by being placed in an environment. Likewise, the environment is not affected or altered as a result of the introduction of the implant. The implant does not affect the environment and the environment does not affect the implant.

A substance is bioactive if the material interacts with the host environment (Dictionary.com 2015). Recent discoveries have resulted in the creation of a synthetic tissue. This tissue is a polymer-based matter that may be layered with the active tissue encouraging growth in damaged areas. An implanted object need not comply with all of the established conditions: nontoxic, bio-inert, bioactive, and biodegradable. However, the developed implanted material is designed to accomplish a certain task, such as a hip replacement, a heart valve, a tooth, or synthetic skin. As long as the implant is not injurious to the host, the product selected is considered a suitable choice. Biodegradable material is a substance that is capable of being decomposed by biological agents, especially bacteria. Dependent of the intent of the implant, the material being used may degrade rapidly or over an extended period of time. If stitches are used internally, the objective of the stitch is to hold the incision closed until healed. After this time the sutures are no longer needed and may decompose. However, a heart valve's decomposition should be over an extended period of time. The individual may pass away before the valve decomposes, but the valve cannot decompose first.

Understanding the effect of a material placed in the biological system is essential when developing an implant. First condition is that the artificial implant is biocompatible with the environment it is being placed. Other factors of interest are the material would not corrode. Corrosion in the host environment may lead to a toxic situation. The material would have a "wear resistance"; sustain the rigors of frictional movement. The implant is appropriate for its replacement in size, weight, and shape. Using a material that deteriorates readily could result in an unsuccessful implant unless the designed is a temporary fix allowing a part of the biological structure to mature or repair itself. An example of this would be an artificial heart being placed into the body while the patient waits of a donor. The material may need to be a low-resistant material, a material that is less apt to deteriorate as a result of frictional wear. Initially the hip replacement was a metallic material, and the hip's ball was placed into the natural socket of the body. Unfortunately the metal caused a rubbing in the socket causing the socket to deteriorate. Types of metal used as an implant need to be scrutinized thoroughly prior to its placement in a host environment if a mechanical operation, the mechanical movement, takes place.

Ceramic material has been found to bond well with bone tissue. It is for this reason that ceramic parts are used in hip replacement, both the ball and socket. A particular type of ceramic material, alumina, is used in implants because of its corrosive resistance and high strength and that it is biocompatible. Alumina is "any of several forms of aluminum oxide used in aluminum production and in abrasives, refractories, ceramics, and electrical insulation" (Dictionary.com 2015).

Endoscopic surgery is becoming more common because this type of surgery is less evasive. Endoscope is a slender, tubular optical instrument used as a viewing system for examining an inner part of the body and, with an attached instrument, for biopsy or surgery (Dictionary.com 2015). A fiber optic light is passed through the tube to view various organs and parts of the body. This allows the surgeon to see organs. Because of the small incision, recovery time is short. Fiber optics has been used in surgery since the 1960s when the optic fiber was bundled for illumination and used with endoscopes for imagining. It wasn't until 1982 that fiber optic used was viewed as providing a means for less invasive surgery. Today, fiber optics has expanded in use beyond mere light imaging and viewing. Special fiber optic fibers were used as sensors measuring temperature, pressure, oxygen, and applied forces. "Fiber optic sensors are seen as safe, reliable, highly stable, biocompatible tools for health-monitoring systems, and they are amenable to sterilization and autoclaving" (Rehman n.d.).

Glass is an inert material resistant to most chemicals. Fiber optics is frequently used due to its small size, and this enables the miniaturization to be used in the surgical field reducing recovery time.

The concept behind fiber optics is simple. It is guiding light using refraction. Refraction is "the alteration of the direction of a ray of light, sound, heat, or the like, in passing obliquely from one medium into another in which its wave velocity is different" (Dictionary.com). The phenomena was first observed in 1840. "Daniel Colladon first described this 'light fountain' or 'light pipe' in an 1842 article titled *On the reflections of a ray of light inside a parabolic liquid stream*. This particular illustration comes from a later article by Colladon, in 1884" ("Optical fiber"). In the 1880s, Alexander Graham Bell and Sumner Tainter invented a photophone in Washington DC. The transmitted sound was carried over an optical beam ("Optical fiber"). Unfortunately, the sound transmitted from the photophone was subject to atmospheric interference and impractical until the fiber optic system was developed. An optimum fiber optic data transmission system was not developed until 1960 when a patent was applied for this process by a German physicist Manfred Borner. The application of the fiber optic patent was not successful due to impurities found in the optic fiber. The material needed to create a pure fiber optic, silica glass, was discovered by Charles Kao in 2006, earning Kao the Nobel Prize in Physics. The advantage for using the fiber optic system is the cost is cheaper than a copper system and the interference is less than copper. Fiber optics does not succumb to electromagnetic interference, an issue the copper systems is subject. There is no environmental noise to be experienced which a copper system has, and the rate of degradation is less than that of copper.

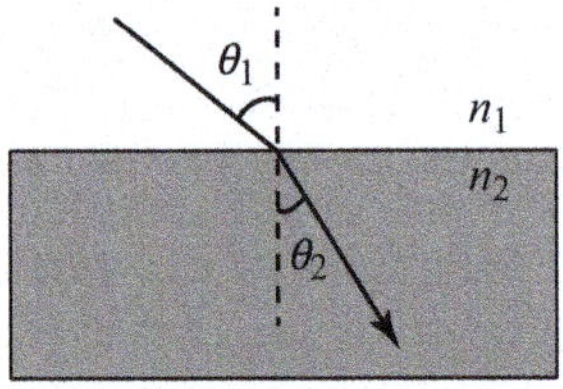

Fig. 12.1 Refraction, illustrating the light transition through different mediums

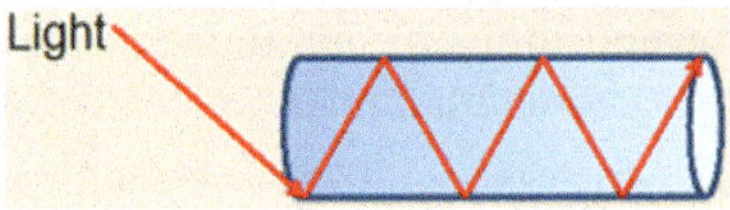

Fig. 12.2 Total internal deflection, Illustrating the behavior of light in a fiber optic material

A discussion of Snell's Law applies a greater understanding of what occurs in a fiber optic wire. When light travels from a less dense medium to a more dense medium, the speed of light is reduced (Fig. 12.1). If the light is moving at an angle as it enters a denser medium, the light is bent toward the normal. The normal is a line measured perpendicular to the plane of the surface at the point of entry. If the light is being transmitted from one medium to another where the transmitting medium is denser than the medium of entry, the light beam is bent away from the normal. The angle of transmission may be wide enough that the light beam does not penetrate the less dense medium. Consequently the light is reflected back into the environment of transmission. The minimum angle where the light does not transmit through the less dense medium is referred to as the critical angle. Total reflection occurs when the angle of transmission is beyond the critical angle. This occurrence is precisely what happens within the fiber optic cable (Fig. 12.2).

The light enters the tube and is reflected off the side walls and continues through the tube. The light is received at the end of the cable and decoded to obtain the original formation (Industrial Physicist 2002).

References

Baylan, M. (2012, January 30). Dental Implants: An Integrative Perspective. Retrieved February 26, 2015, from http://www.westonaprice.org/holistic-healthcare/dental-implants-an-intergrative-perspective

Dictionary.com (n.d.). Retrieved March 3, 2015, from http://dictionary.reference.com/

Ganapathi, M. (2010, January 1). Does size of the ball matter in THR? Retrieved March 2, 2015, from http://www.ganapathihipkneesurgeon.com/does size of the ball matter in hip replacement.html

Hip Replacement. (2013, July 1). Retrieved February 26, 2015, from http://www.niams.nih.gov/Health_Info/Hip_Replacement/default.asp

Industrial Physicist. (2002, March 1).

Jacobson, J. (2013, September 25). Hip Replacement Imagining. Retrieved February 26, 2015, from http://www.mdguidelines.com/complications-of-devices-implants-and-grafts

Rehman, S. (n.d.). Specialty Optical Fibers Make Surgery Less Invasive. Retrieved February 26, 2015, from http://www.photonics.com/Article.aspx?AID=48684

Chapter 13
Polymers

In the movie the Graduate, Dustin Hoffman was informed to that the future will be based on plastics. That was nearly 50 years ago, and today we see plastics all around the world. The automotive industry was able to increase the efficiency of the car by reducing its weight. This is accomplished by the introduction of plastics or more specifically polymers. However, as the polymer increased in popularity, a problem arose, what is to be done with the waste products or refuge because the polymer material is not biodegradable. A better understanding of the material's structure will allow for a better awareness of what may need to be accomplished to resolve the excessive tonnage of waste products in using polymer as well as a better understanding on how to use this material.

Polymers

A polymer is a substance made up of a partial molecular structure or entirely of a large chain of bonded materials forming plastics or resins. "The properties of anything made out of polymers really reflect what's going on at the ultra-tiny (molecular) level. So, things that are made of polymers look, feel, and act depending on how their atoms and molecules are connected, as well as which ones we use to begin with! Some are rubbery, like a bouncy ball, some are sticky and gooey, and some are hard and tough, like a skateboard." (http://www.pslc.ws/macrog/kidsmac/basics.htm)

R. A. Wesolowski et al., *The World of Materials*,
https://doi.org/10.1007/978-3-030-17847-5_13

Back in 1967, there was a movie entitled *The Graduate*. The movie was directed by Mike Nichols and starred Dustin Hoffman and Anne Bancroft. During the film, Walter Brooke, playing the part of Mr. McGuire, made mention to Benjamin (Dustin Hoffman) to remember plastics, the future (Quotes n.d.). Plastics are polymers. Polymers are defined as something that is made of many units. To make the chain, many links or "mers" are connected together or polymerized. This can be visualized as a connection of paperclips, making a chain (Fig. 13.1).

Polymers have been in existence since the beginning of time. A tortoise shell and animal horns are examples of a polymer. Tar, shellac, and tree sap, used to produce latex, are also examples of polymers. Many of these polymers were processed under pressure and heat to form hair ornaments and jewelry.

Many common polymers are found to be hydrocarbons. These are small units bonded together to form large chains. A polymer chain is illustrated below (Fig. 13.2).

Polypropylene, polybutylene, polystyrene, and polymethylpentene are polymers that contain only hydrogen and carbon elements. Although carbon and hydrogen are the base elements, other elements may be added to create a unique polymer. Polyvinyl chloride (PVC) contains chlorine. Nylon contains nitrogen and oxygen. Teflon contains fluorine. Polyester and polycarbonates contain oxygen. There are some polymers that do not use a carbon base as their principal element. One such polymer, having silicon as its primary ingredient, is found in the product called "Silly Putty."

Consider how spaghetti appears on a plate. This illustration resembles the arrangement of a polymer that is amorphous. The amorphous polymer does not have a long arrangement, and this amorphous structure generally results in a transparent

Fig. 13.1 Paper clip chain, www.thriftyfun.com 300 × 200 Search by image

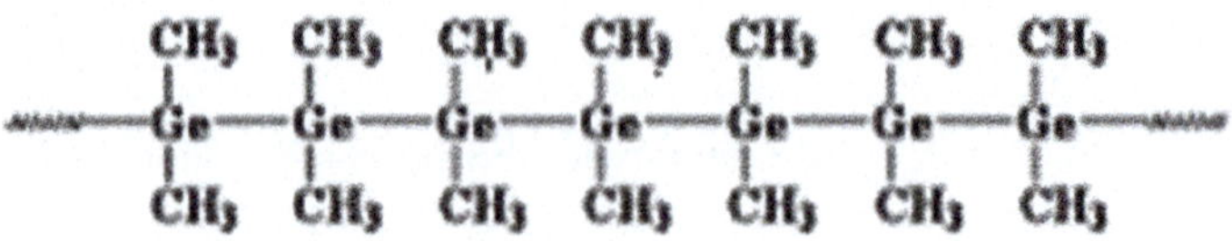

Fig. 13.2 Polydimethylgermane, illustrating the pattern of a polymer chain

substance. An example of this type of polymer would be PlexiglasTM, headlights, and contact lenses. Not all polymers are transparent. There are polymers that are translucent and opaque. The opaque or translucent polymers are usually crystalline in structure. Amorphous materials are a result of rapid cooling, quenching. A controlled cooling process, such as annealing, renders a controlled atomic structure which is more organized making a crystalline polymer. Less light will pass through material containing a great amount of crystalline polymer. Engineers can manipulate the structure of the polymer that can result in a more refined final product.

Polymers can be grouped into two major areas, thermoset and thermoplastic. Most plastics are thermoplastic. This polymer once created can be reheated and then reformed. This can be done repeatedly. A thermoset polymer cannot be reheated. If reheating is attempted, the plastic will be scorched and burned.

Considering that chemicals can be contained in plastic containers implies that chemicals have a minimal effect on a polymer. Polymers are used in insulating, in microwavable materials, and in electrical outlets in the environment. A skier uses polymer clothing to maintain body heat. The density of a polymer-based material is low. The material is lightweight and used when weight constraints are necessary. The application varies from dime store toys to the space program, where it serves as the structure for a space station. A polymer, possessing modeling properties, can be added to other products as an additive, allowing to the product receiving the polymer to assume new characteristics. Polymers may be added to adhesives or paints to prevent them from dripping.

A key issue with plastic use is its proneness to deteriorate. Although plastics are prone to deteriorate, they do not decompose. The thermoplastic polymer can be recycled due to the fact that polymers can be reheated and reformed. Plastics can be remolded into picnic tables, fences, and outside toys. Pyrolysis is the process of applying heat and pressure to polymers and reversing the process of its creation and returning the material into an oil base. This process requires no oxygen. This artificial process mimics the natural process of decomposition which would need to span many years.

A hydrogen carbon bond is a covalent bond. Hydrogen, having one electron in its valence shell, and carbon, having four electrons in its valence shell, allow one electron to be shared between the carbon and the hydrogen element, stabilizing the hydrogen. Unfortunately, this sharing of the single electron, between the hydrogen and the carbon, does not stabilize the carbon. There is a need of three more electrons to stabilize the outer valence shell (Fig. 13.3).

The carbon-hydrogen chain can be expanded making a long chain, or becoming polymerized (Fig. 13.4).

At times, a stronger polymers is not needed; rather flexibility is required. The flexibility must be so designed that the polymerization chain does not become tangled. To accomplish this, plasticizers are added. Plasticizers or dispersants are additives that increase the plasticity or fluidity to a material. These can include plastics, cement, concrete, wallboard, and clay. Plasticizers prevent the chains from lying too close together preventing them from crystallizing, thus making them more flexible (Fig. 13.5).

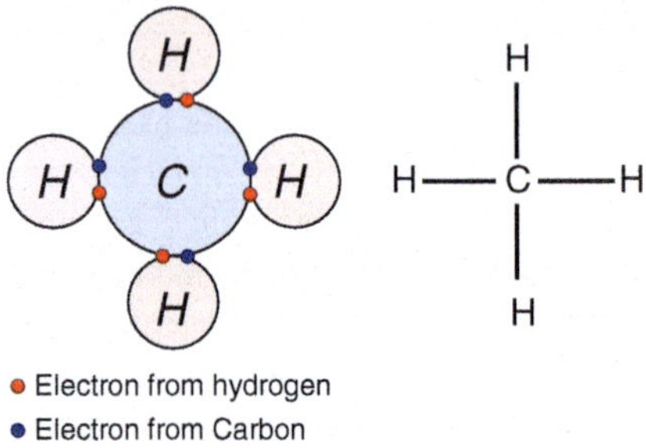

Fig. 13.3 Carbon hydrogen chain, showing the chemical composition of the mixture

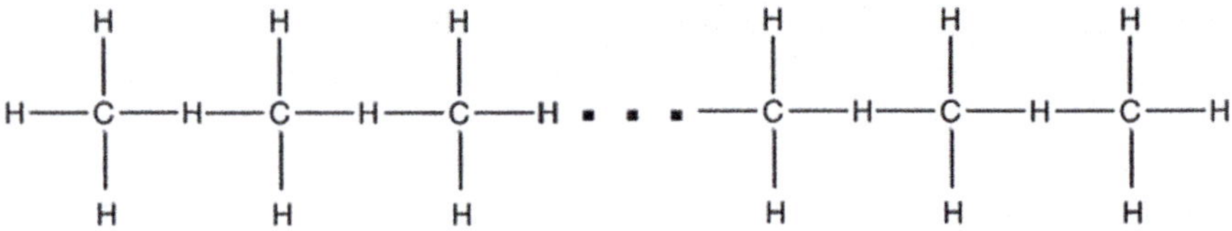

Fig. 13.4 Long-chain illustration of a carbon-hydrogen chain

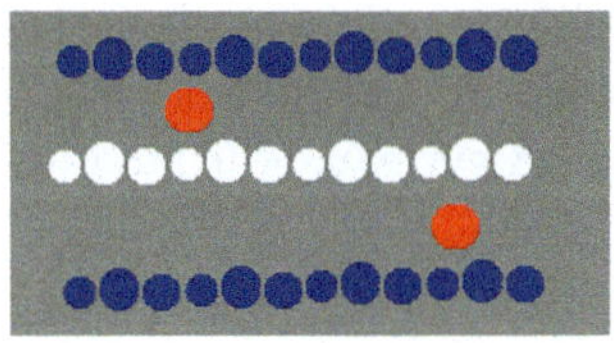

Fig. 13.5 Plasticizers, illustrating where plasticizers are placed in a chain

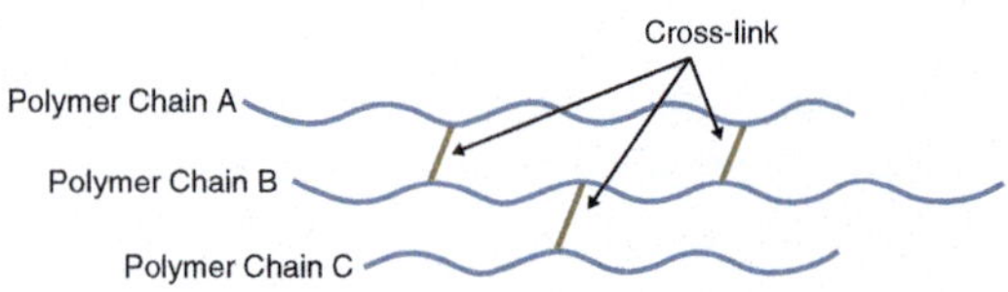

Fig. 13.6 Crosslinks, illustrating where a crosslink is located in a chain

To make the polymer stronger, a "cross-link" is introduced. The "cross-link" is a bond joining one polymer to another (Fig. 13.6). The bond could be covalent or ionic. In the synthetic world of polymers, the cross-linking supports a different property of the polymer. In the biological science, cross-linking refers to a probed investigation of a protein-protein interaction. When cross-linking is added to long rubber molecule, the linking stops, the flexibility is reduced, and hardness increases. Cross-links can be formed through chemical reactions that are initiated by heat, pressure, change in pH, or radiation. The chemical process of vulcanization is a form of cross-linking that changes rubber to the hard, durable material associated with car and bike tires.

Table 13.1 Features of alkanes of different sizes, self-drawn

Number of carbon atoms	Feature	Name	Example	Scale length	Melting point in Celsius
1	Smallest hydrocarbon	Methane	Natural gas	0.3 cm	−183
6	Smallest liquid hydrocarbon	Hexane	Charcoal lighter fluid	2 cm	95
18	Smallest solid hydrocarbon	Octadecane	Very soft paraffin wax	5 cm	28
100	A "waxy" poly(ethylene)	Decadecane		30 cm	110
1000	A small poly(ethylene)		Poly(ethylene)	3 m	Softens at 130
10,000	Typical poly(ethylene)		Poly(ethylene)	30 m	Softens at 150

Adapted from Lynn Higgins 708.447.6339 Iynhiggins@aol.com
Alkanes are organic compounds that consist entirely of single-bonded carbon and hydrogen atoms and lack any other functional groups

The polymer can be natural or synthetic. A natural polymer is found in plants and animals. One such polymer is starch which is polymer of glucose and is a food reserve for plants. Another polymer of glucose is cellulose which serves as structural material of plants. This cellulose is made by the plant with the glucose during photosynthesis. Polymers of proteins, *a*-amino acids, the second largest building block next to water, of human muscle, cells, and other tissues, is a combination of hydrogen, oxygen, carbon, and nitrogen (other elements may be found), ("Amino acids,"). Nucleic acids are members of polymers that are poly-molecules or large biomolecules. Nucleic acids, which include DNA (deoxyribonucleic acid) and RNA (ribonucleic acid), are made from monomers known as nucleotides. Polymers such as polysaccharides (starches, cellulose), proteins, and nucleic acids, which control various life processes of plants and animals, are called biopolymers.

Polymers produced in a laboratory are called synthetic polymers. These man-made polymers are referred to as Teflon, polyethylene, PVC nylon, and synthetic rubber. Chemically modified polymers from the natural process are referred to as semisynthetic polymers. A viscose rayon, the first man-made fiber, is not synthetic. It is made from wood pulp, a cellulose-based material in long fibers (Table 13.1).

Reference

Quotes. (n.d.). Retrieved March 3, 2015, from http://www.imdb.com/title/tt0061722/quotes

Chapter 14
Pros and Cons of Plastics

What do telephones, carpets, and even fleece we wear have in common? (Fig. 14.1) They all contain plastic. We store our food in it, drink water from it, and even brush our teeth with it. It's cheap, convenient, and versatile – but how much do we really know about this wonder material? (Fig. 14.2).

Being able to tote around a bottle of water or soda and discard it when it's empty is undeniably convenient. Water fountains are occasionally soiled or in disrepair and aren't available anywhere, anytime. Plastic bottles are generally cleaner than reusable bottles. Because they're intended for one-time use, they don't need to be maintained or washed (Fig. 14.3).

However, each year, more than 26 billion plastic water bottles are thrown away, and of those, less than 15% are recycled. The bottles that aren't recycled end up in landfills and don't degrade. The production of plastic bottles is incredibly taxing on resources, including water. Transporting and distributing water bottles also strains gas, oil, and other natural resources (Fig. 14.4).

It lasts a long time! Consider the life span of these popular plastic products (Fig. 14.5):

- Plastic water bottle – 450 years
- Disposable diapers – 500 years
- Plastic 6-pack collar – 450 years
- Extruded polystyrene foam – over 5000 years (Fig. 14.4)

Is longevity a pro or a con? (Fig. 14.6).

Plastics are lightweight! Oil and natural gas are the major raw materials used to manufacture plastics. The plastic production process often begins by treating components of crude oil or natural gas in a "cracking process." This process results in the conversion of these components into hydrocarbon monomers such as ethylene and propylene. Further processing leads to a wider range of monomers such as styrene, vinyl chloride, ethylene glycol, and many others. These monomers are then

From presentation at NSTA Conference March 16, 2018, 8 AM.

R. A. Wesolowski et al., *The World of Materials*,
https://doi.org/10.1007/978-3-030-17847-5_14

Fig. 14.1 Collection of polymer products

Fig. 14.2 Polymers as a betterment

chemically bonded into chains called polymers. The different combinations of monomers yield plastics with a wide range of properties and characteristics (Fig. 14.7).

Some polymers float in water while others sink. But, compared to the density of stone, concrete, steel, copper, or aluminum, all plastics are lightweight materials (Fig. 14.8).

Given objects. Metal and polymers as well having determined the volume by utilizing an overflow technique, the volume was determined to be:

Massing the objects, the object's mass was determined to be (Fig. 14.9):

Fig. 14.3 Plastics as refuge

Fig. 14.4 Longevity of polymers

The activity of determining the density allows the students to understand that not all polymers are the same, and yet they remain less dense than metals of similar volumes (Fig. 14.10).

Temperature change. The polymers we work with follow the same principles as in the human body (Fig. 14.11).

The hotter the environment becomes, the less performance we can expect. Engineers classically are trained in materials; they have knowledge, such as metals (Fig. 14.12).

Consequently, they are accustomed to a level of certainty where *temperature-dependent behavior* is concerned and *that polymers do not provide*. Even a relatively low-performing metal such as aluminum has essentially the same mechanical properties between room temperature and 250–300 °C, and even when approaching the melting point, retention can be as high as 80% of room temperature performance

Fig. 14.5 Lengths of time of polymer existence

Fig. 14.6 Good reason for polymers

(Fig. 14.13). This level of consistency is even better for materials like copper, brass, and steel, where the melting points are higher. This certainty is aided by the fact that the building blocks for metallic substances are small and readily organize into a well-defined and predictable crystal structure.

That is not the case for polymers (Fig. 14.14).

The individual molecules that make up polymeric materials are very large and have an extended chain-like shape that results in an entangled structure (Fig. 14.15).

This entanglement is beneficial in some respects. The relatively high levels of elongation that most polymers exhibit without breaking are due in large part to chain entanglement. However, this entanglement also restricts the freedom required

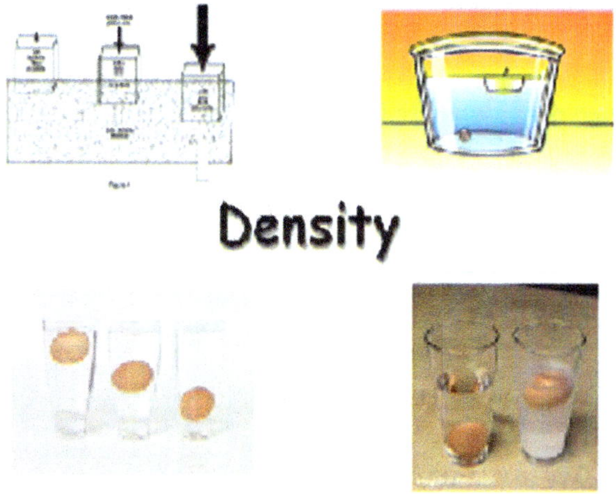

Fig. 14.7 Density of materials

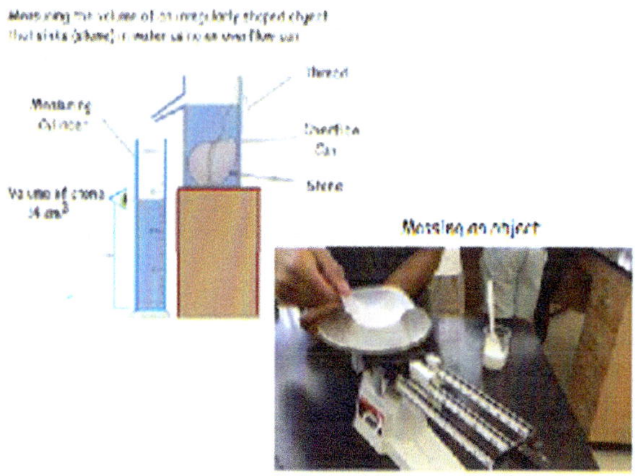

Fig. 14.8 Determining density

at a molecular level to organize into crystals. Consequently, no polymer under normal processing conditions is fully crystalline, and some polymers do not crystallize to any significant degree.

This lack of a predictable and repeatable structure gives rise to a situation where changes in temperature always influence the mechanical properties of these materials (Fig. 14.16).

This phenomenon can be captured by performing mechanical tests at various temperatures on any given material.

This technique can measure so many aspects of polymer behavior that whole books have been devoted to the topic. However, for our purposes we can focus on one small aspect: the ability to measure the elastic modulus of a material as a function of temperature.

How does temperature affect these plastics?

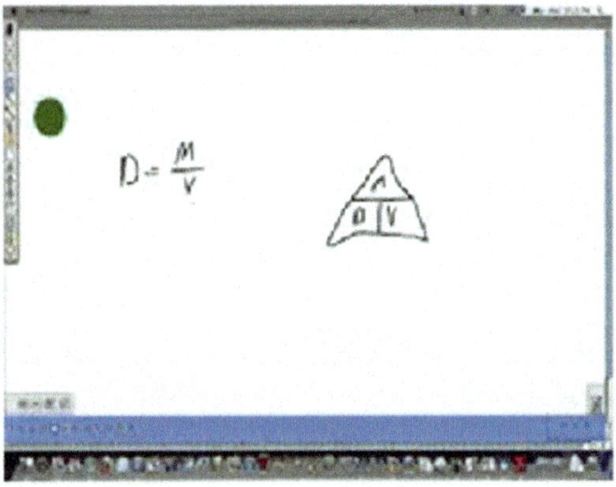

Fig. 14.9 Determining density

Fig. 14.10 Temperature and polymers

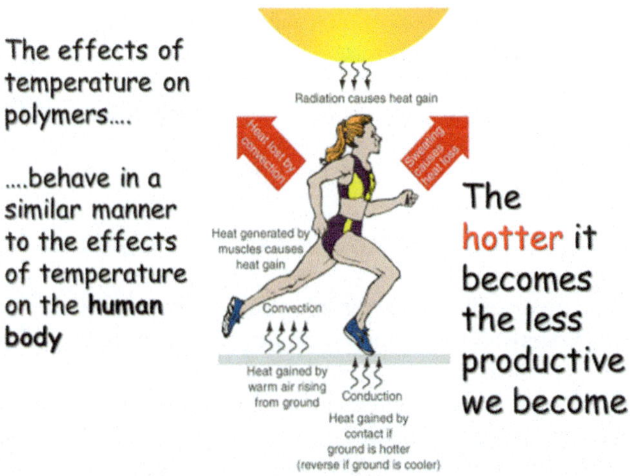

Fig. 14.11 Change of temperature on the human body

Fig. 14.12 Further effects of temperature

Low performing metals

Table 1. Material properties of aluminum alloy 6061-T6, alumina and steel.

Material	Temperature (°C)	Thermal conductivity (W.m⁻¹.K⁻¹)	Heat capacity (J.kg⁻¹.K⁻¹)	Density (kg.m⁻³)	Thermal expansion (×10⁻⁶ K⁻¹)	Young's modulus (GPa)	Yield stress (MPa)	Poisson's Ratio	Melting point (°C)
Aluminum alloy 6061-T6	0	162	917	2703	22.4	69.7	277.7	0.33	582-652
	98	177	978	2685	24.61	66.2	264.6		
	201	192	1028	2657	26.6	59.2	218.6		
	316	207	1078	2630	27.6	47.78	66.2		
	428	223	1133	2602	29.6	31.72	17.9		
	571	253	1230	2574	34.2	0	0		
Alumina	25-727	16.7	950	3990	7.5	300	-	0.2	2127
Steel	0	34.5	470	7800	10.1-16.6	68.9-207		0.21-0.3	1210-1530
	98	34.5	485						
	201	33.8	520						
	316	31	560						
	428	28.5	620						
	571	26.8	700						
	650	25.8	760						

Fig. 14.13 Temperature on metals

Place three (3) near identical samples into three unique temperature environments: (1) room temperature, (2) freezing (freezer unit), and (3) boiling water (Figs. 14.17 and 14.18).

The materials were then placed on a tensile tester, and the elasticity, plasticity, and the deformation were recorded and the data analyzed. The graphic results are below (Fig. 14.19):

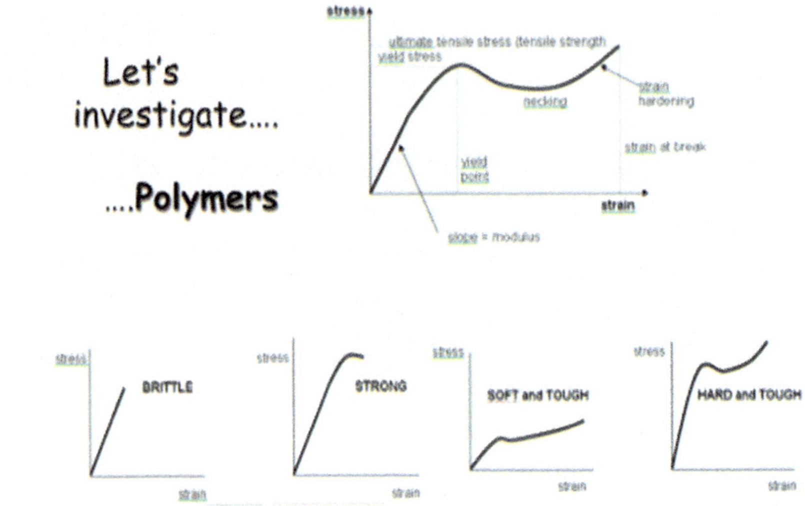

Fig. 14.14 Effects of temperature on polymers

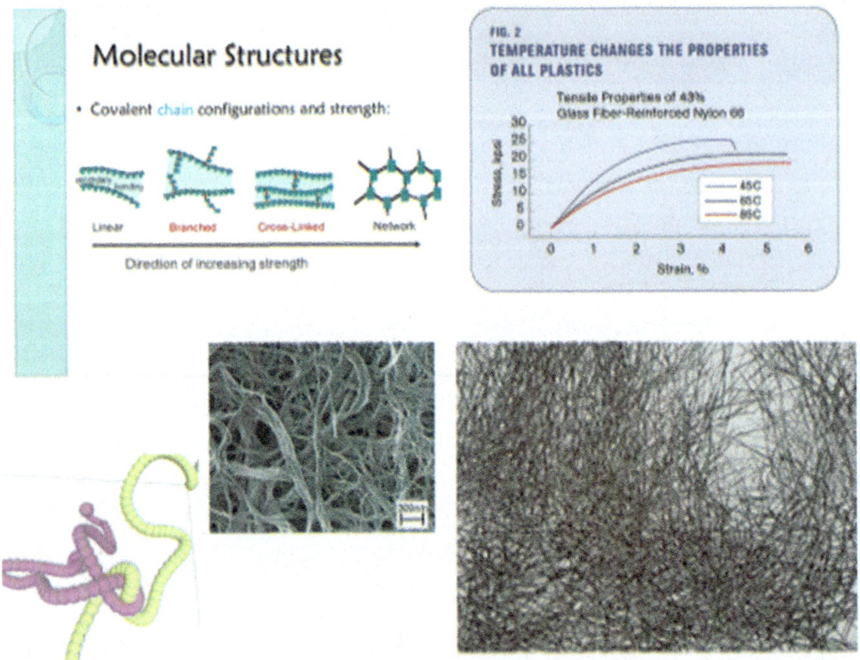

Fig. 14.15 Molecular properties and entanglement

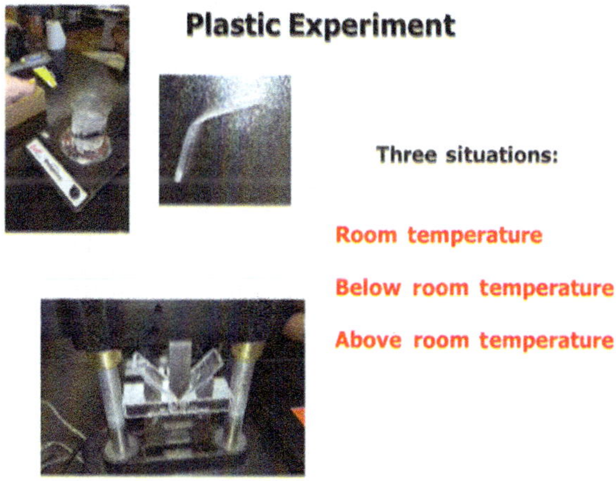

Fig. 14.16 Polymer experiment

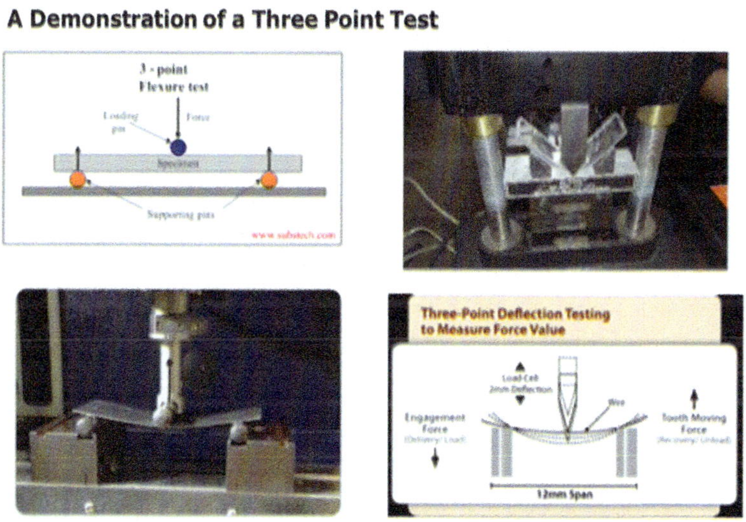

Fig. 14.17 Three point test

Room temperature showed a normal expectation (Fig. 14.20).

There was a gradual increase in slope during the elasticity state, transitioning into a plasticity state – notice the curvature which illustrates the transitioning. Finally observe the abrupt deformation that occurs and remember the temperature to which the event happens (Fig. 14.21).

From the data chart, notice the difference between the elastic and the plastic constant (Fig. 14.22).

Room Temperature:

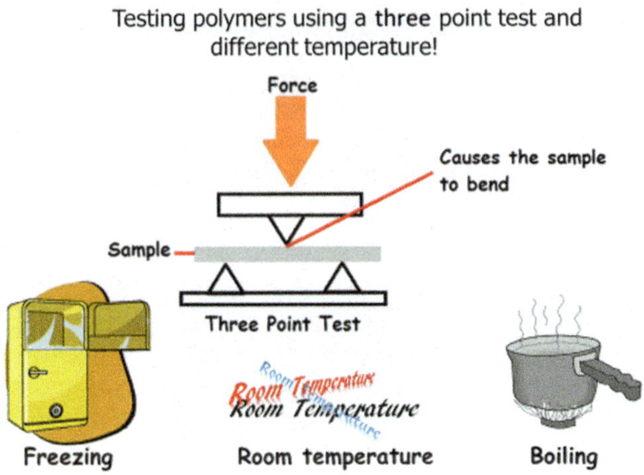

Fig. 14.18 Three point test

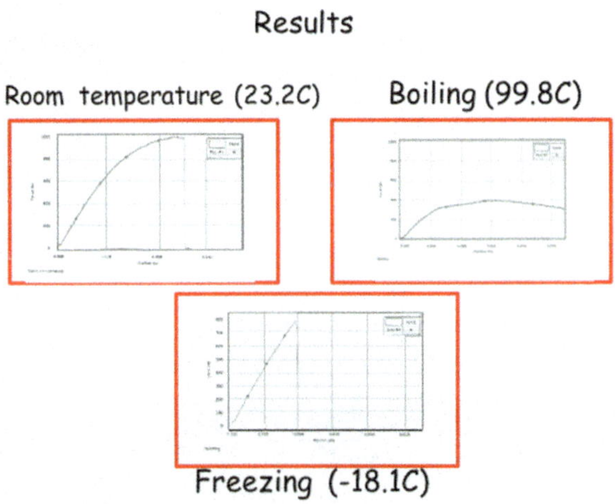

Fig. 14.19 Tensile test results

Elastic – 1.57×10^5 N/m
Plastic – 3.7×10^4 N/m

When an exterior stress is applied to a solid body, the body tends to pull itself apart. This causes the distance between atoms in the lattice to increase. Each atom tries to pull its neighbor as close as possible. This causes a force trying to resist the deformation. This force is known as strain. The elastic deformation of a solid is a reversible process; when the applied stress is removed, the solid returns to its original state.

Plasticity is the property that causes irreversible deformations on an object or a system. Such deformations can be caused by forces and impact (Fig. 14.23).

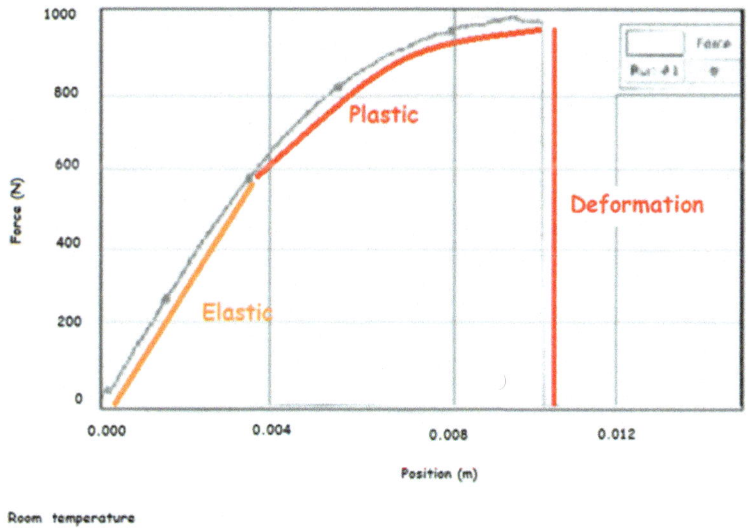

Fig. 14.20 Tensile test at room temperature

Fig. 14.21 Tensile test data

Room Temperature

Polymer and Temperature
Three Point Test

Three Point Test – Room Temp Temperature 23.2 C

Elasticity Force	580 N
Distance Displaced	0.0037 m
Plasticity	(940 – 580 = 360) N
Distance Displaced	(0.0103 – 0.0037 = 0.00966 m
Deformation	940 N

Elastic constant 580 N / 0.0037 m) = 1.57 x 10^5 N/m

Plastic Constant 360 N / 0.00966 m = 3.7 x 10^4 N/m

After the ultimate strength, the material begins to “neck” making unevenness of the density over length. This makes very low-density areas in the material making it easily breakable (Fig. 14.24).

At boiling, the elastic state is transitioned into the plastic state and extending over time. This illustrates the flexibility of the polymer material at high temperatures; thus the deformation occurs only after a substantial amount of time (Fig. 14.25).

Fig. 14.22 Reduction in plasticity versus elasticity

Notice there is a reduction in the constants:

Room Temperature:

Elastic – 1.57 x 10^5 N/m

Plastic - 3.7 x 10 ^4 N/m

When an exterior stress is applied to a solid body, the body tends to pull itself apart. This causes the distance between atoms in the lattice to increase. Each atom tries to pull its neighbor as close as possible. This causes a force trying to resist the deformation. This force is known as strain. The elastic deformation of a solid is a reversible process, when the applied stress is removed the solid returns to its original state.

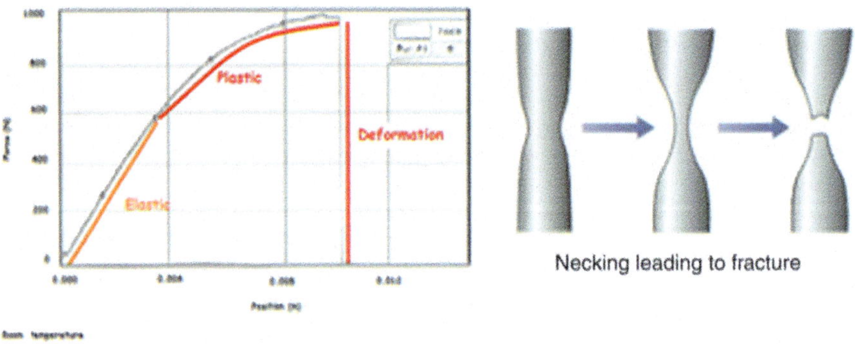

Fig. 14.23 Necking leading to a fraction

Fig. 14.24 Tensile results of heated polymer

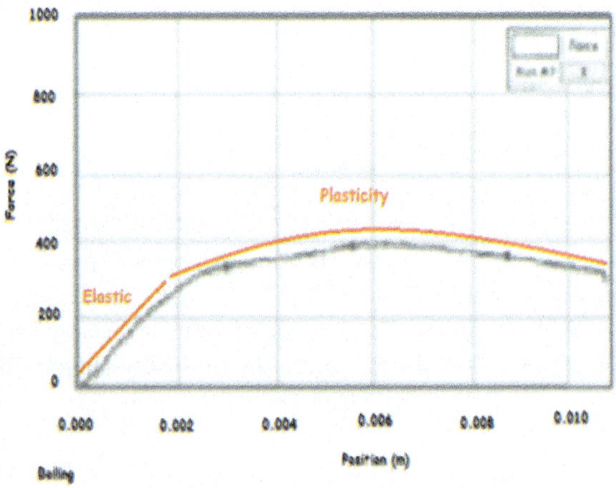

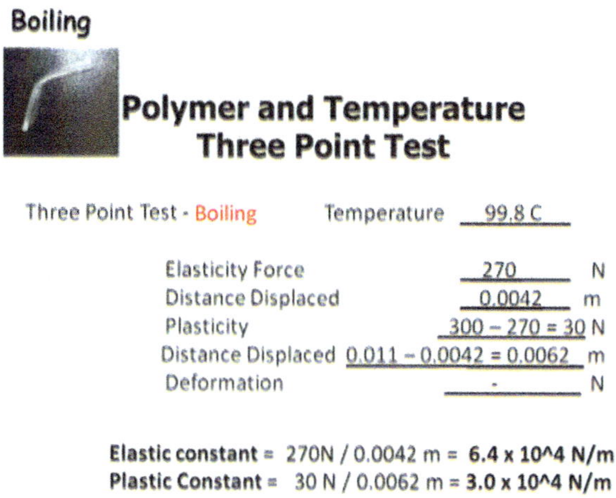

Fig. 14.25 Data from heated tensile test

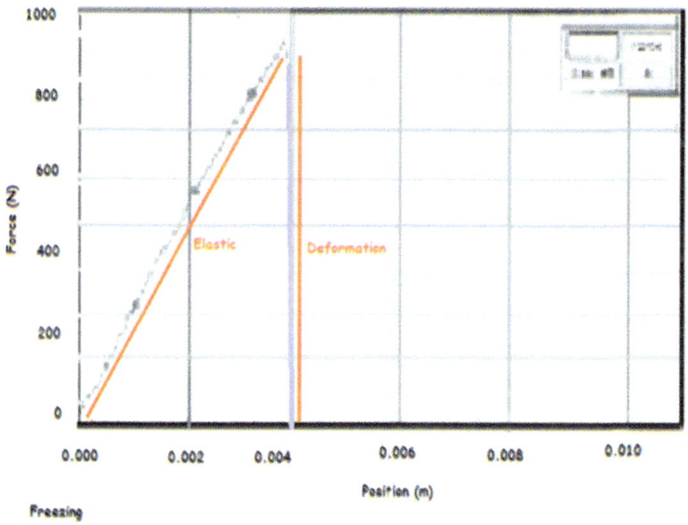

Fig. 14.26 Polymers at freezing point

The constants for elasticity versus plasticity are almost identical as seen by the slopes of the graph (Fig. 14.26).

At the freezing point, the slope is more extreme, with little to no plasticity occurring before the deformation occurs. This illustrates brittleness as a result to extreme reduction in temperatures (Fig. 14.27).

Having removed energy, freezing, the material becomes more rigid and less malleable. The material itself becomes brittle, acting as a ceramic. As stress is applied to the colder material, the material undergoes an elastic change until fracturing occurs.

Freezing

Polymer and Temperature
Three Point Test

Three Point Test - Freezing Temperature -18.1C

Elasticity Force	880	N
Distance Displaced	0.0042	m
Plasticity	-	N
Distance Displaced	-	m
Deformation	880	N

Elastic constant = 880N / 0.0042 m = 2.1 x 10^5 N/m

Fig. 14.27 Data of freezing point

Notice that the elastic constant of the colder material is higher than the plastic at room temperature.

Elastic – 1.57×10^5 N/m room temperature
Elastic – 2.1×10^5 N/m freezing

This indicates the greater rigidity in the colder plastic.

Presentation on the comparison of the freezing, boiling, and the normalcy of temperature on polymers. Relate this to the 1986 Challenger disaster (Fig. 14.28).

In 1986 the Challenger exploded due to an O-ring problem. Temperature was cold that day; the material of the O-ring was subjected to this environmental influence. They became more brittle than the norm, that is, less pliable, less flexible. The results rendered the death of the entire crew, including a teacher, Christa McAuliffe (Fig. 14.29).

Strength is a mechanical property that you should be able to relate to, but you might not know exactly what we mean by the word "strong" when we're talking about polymers. First, there is more than one kind of strength. There is *tensile* strength. A polymer has tensile strength if it is strong when one pulls on it like this:

Tensile strength is important for a material that is going to be stretched or under tension. Fibers need good tensile strength.

Elastomers are polymers that are able to stretch a long distance and still bounce back. Most of them can stretch from 500% to 1000% elongation and return to their original lengths without any trouble.

Then there is *compressional* strength. A polymer sample has compressional strength if it is strong when one tries to compress it, like this:

Concrete is an example of a material with good compressional strength. Anything that has to support weight from underneath has to have good compressional strength.

Fig. 14.28 Challenger disaster in 1986

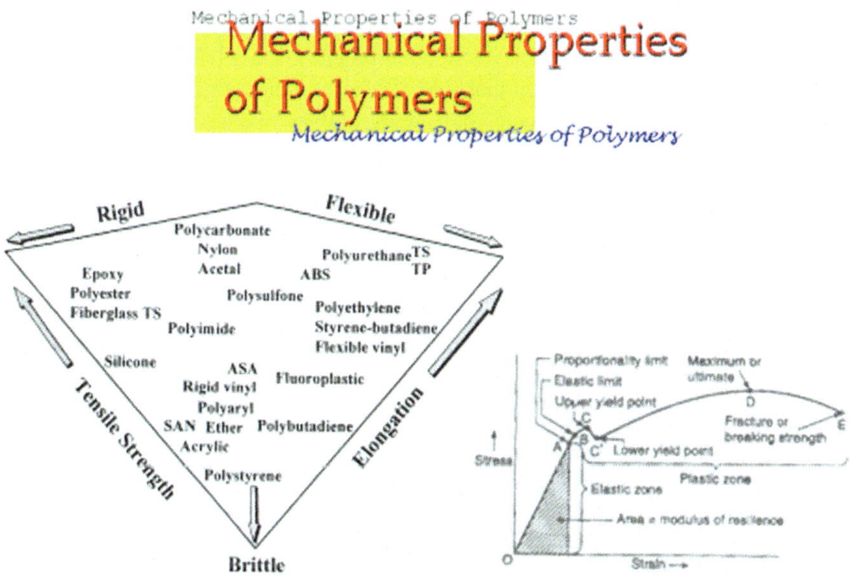

Fig. 14.29 Mechanical properties of polymers

There is also *flexural* strength. A polymer sample has flexural strength if it is strong when one tries to bend it, like this:

A sample of *torsional* strength if it is strong when one tries to twist it. Then there is *impact* strength. A sample has impact strength if it is strong when one hits it sharply and suddenly, as with a hammer (Fig. 14.30).

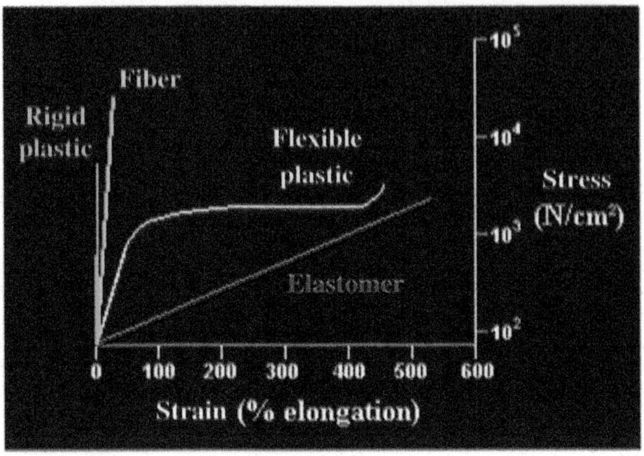

Fig. 14.30 Graphic analysis. (After Odian G. Principles of polymerization. 3rd ed. New York: Wiley; 1991. p. 34)

The graph compares typical stress-strain curves for different kinds of polymers. You can see in the green plot that rigid plastics such as polystyrene, polymethyl methacrylate and polycarbonate can withstand a good deal of stress, but they won't withstand much elongation before breaking. There isn't much area under the stress-strain curve at all. So we say that materials like this are strong, but not very tough. Also, the slope of the plot is very steep, which means that it takes a lot of force to deform a rigid plastic. (I suppose that's what it means to be rigid, now doesn't it?) So it's easy to see that rigid plastics have high moduli. In short, rigid plastics tend to be strong, resist deformation, but they tend not to be very tough, that is, they're brittle (Fig. 14.31).

Polyurethane is a foam *insulation* material that contains a low-conductivity gas in its cells. Foil and *plastic* facings on rigid polyurethane foam panels can help stabilize the R-value, slowing down thermal drift (Fig. 14.32).

The thermal conductivity of a material determines how good an insulator it is. The official definition of thermal conductivity is the quantity of heat transmitted through a unit thickness in a direction normal to a surface of unit area, due to a unit temperature gradient under steady-state conditions. A simpler way of saying this is thermal conductivity is the ability to conduct heat. The best materials for insulation are the ones with the lowest thermal conductivity (Fig. 14.33).

Given two identical pieces of metal, butted against each other, measure the thickness, and measure the area. Apply heat to one side. Using an infrared thermometer, measure the temperature on the side the heat is applied and the opposite side of the butted materials.

Repeat the process placing an insulated material between the metals. Again measure the thickness, but the area should be the same. Apply heat and determine the temperatures on both surfaces of the areas.

Fig. 14.31 Polymers as insulators

The Foil or plastic facing help stabilize the R-value

Why???

The thermal conductivity of material determines how good an insulator a material happens to be!

Fig. 14.32 Thermal conductivity

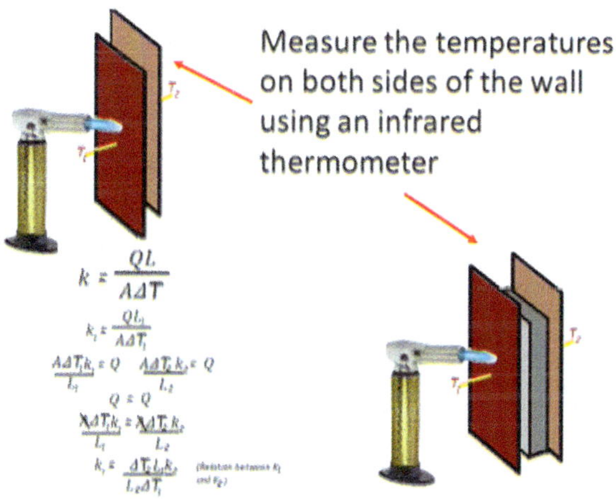

Fig. 14.33 Thermal conductivity test

$$k = \frac{QL}{A\Delta T}$$

$$k_1 = \frac{QL_1}{A\Delta T_1}$$

$$\frac{A\Delta T_1 k_1}{L_1} = Q \quad \frac{A\Delta T_2 k_2}{L_2} = Q$$

$$Q = Q$$

$$\frac{A\Delta T_1 k_1}{L_1} = \frac{A\Delta T_2 k_2}{L_2}$$

$$k_1 = \frac{\Delta T_2 L_1 k_2}{L_2 \Delta T_1} \quad \text{(Relation between } K_1 \text{ and } K_2\text{)}$$

Using the given formula, determine the heat transfer and then the thermal conductivity.

How beneficial was the polymers being used as an insulator? (Investigate the effect of the polymer, Styrofoam, on the transfer of heat. An obvious strength of polymers is in the insulation properties.)

Corrosion on polymers, both plastics and rubber materials, is in many cases similar to metals, but in other cases, it looks very different. Corrosion attacks on polymers are often hard to discover; the material may look normal but can in fact be embrittled and have lost its mechanical strength (Fig. 14.34).

Polymers consist of a network with molecular chains mainly consisting of carbon, hydrogen, and oxygen. Corrosion by chemical reaction changes the configuration of the polymer chains. Using a polymer coating may assist in reducing the degradation of corrosion experienced by the metal, but the polymer itself may begin to deteriorate when UV rays react with the polymer.

With the many advances in the medical field, polymers have served in many capacities (Fig. 14.35).

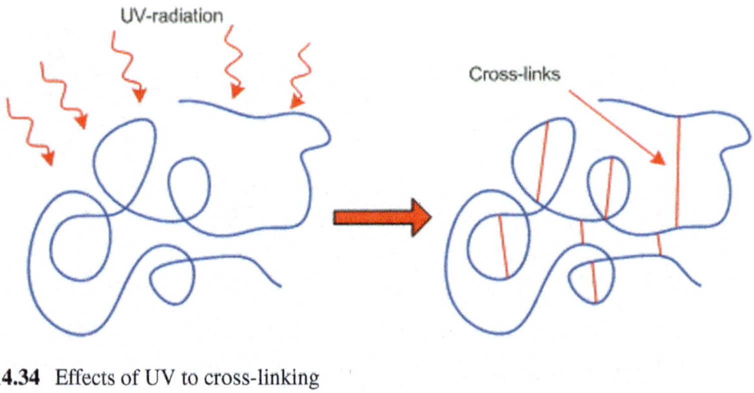

Fig. 14.34 Effects of UV to cross-linking

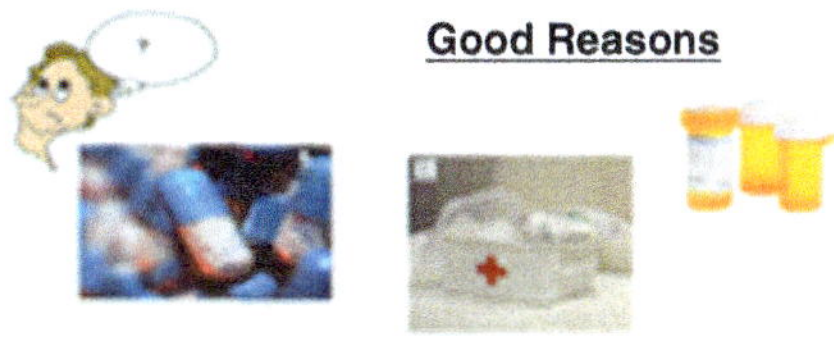

Fig. 14.35 Medical uses of polymers

The pill bottle is an example where plastic is used, the pills themselves are housed in a "polymer" container, and the needles are metal, but the hypodermic is plastic and disposable. The most common polymer used in the medical field is PVC, polyvinyl chloride. Various medical applications are made from, or contain, PVC, including:

1. Containers for blood, urine continence, and ostomy products
2. Containers for intravenous solution giving sets
3. Heart and lung bypass sets
4. Catheters and cannulae
5. Tubing for dialysis, endotracheal, feeding, and pressure monitoring
6. Surgical and examination gloves
7. Inflatable splints
8. Inhalation masks
9. Blister packs for pills and tablets (Fig. 14.36)

Polymer research has led to the discovery of fire proof polymers (Fig. 14.37).

Fire-safe polymers are polymers that are resistant to degradation at high temperatures. There is need for fire-resistant polymers in the construction of small, enclosed spaces such as skyscrapers, boats, and airplane cabins (Fig. 14.38).

Some fire-safe polymers naturally exhibit an intrinsic resistance to decomposition, while others are synthesized by incorporating fire-resistant additives and fillers.

Let's summarize the main reasons for using polymers in the world we live (Figs. 14.39 and 14.40).

1. The main plus is that polymers are cheap to use.
2. Polymers are lighter.
3. Excellent insulators.
4. Convenient.
5. Aid in corrosion abatement.
6. Fire retardant.
7. Purity.
8. Medical profession (Figs. 14.41 and 14.42).

There is quite a bit of it being used!

According to a study by the Ocean Conservatory in 2015, plastic bags came in second to discarded cigarette butts as the most identified type of refuse *aka*

Fig. 14.36 Medical uses

Fig. 14.37 Fire prevention and polymers

Fig. 14.38 Polymer usage and fire prevention

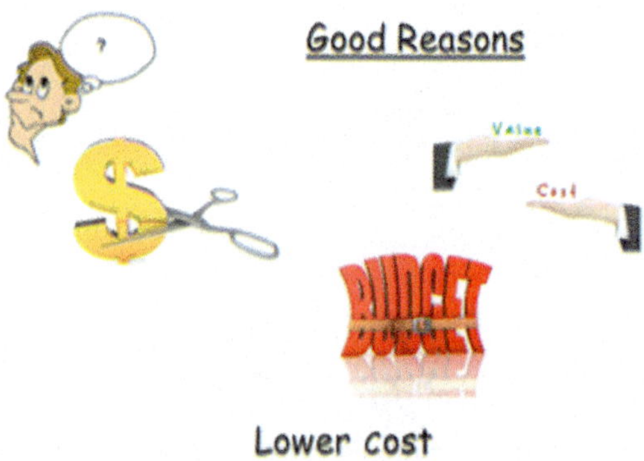

Fig. 14.39 Reasons to use polymers

Fig. 14.40 Reasons to use polymers

Fig. 14.41 Questioning the use of polymers

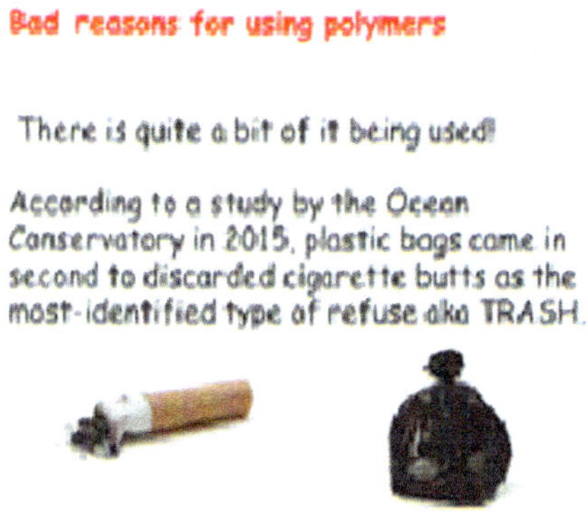

Fig. 14.42 Cons of polymers

TRASH. Synthetic polymers, commonly known as plastics, have been entering the marine environment in quantities paralleling their level of production over the last half century. However, in the last two decades of the twentieth century, the deposition rate accelerated past the rate of production, and plastics are now one of the most common and persistent pollutants in ocean waters and beaches worldwide (Fig. 14.43).

There is excessive litter by us humans, and that litter is *centered around polymers* and in particular plastic bags.

While some of these items are recycled, the growth of plastic consumption and its improper disposal currently outpace efforts to recycle and produce postconsumer plastic materials (Fig. 14.44).

Fig. 14.43 Liter

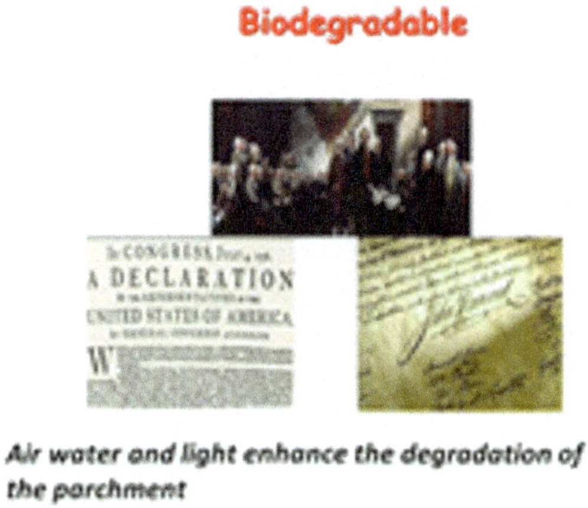

Fig. 14.44 Biodegradable

Fig. 14.45 Degradation of materials

For the most part, decomposition usually implies too many biodegradable. *Biodegradable* means "capable of being decomposed by bacteria or other living organisms." When landfills are considered, the natural reaction is that the waste will decompose as a result of rot or bacterial deterioration (Fig. 14.45).

When considering a plastic, or a polymer, the natural perception is that the material will last for a very long time. The plastic or polymer is not viewed as being able to decompose via a mean such as found in a typical dump.

Biodegradation requires air water or light to enhance the process. Polymers are usually buried in a land heap with soil, removing light, an essential need in the breakdown of plastics (Fig. 14.46).

Some plastics are engineered to decompose ...

...... many do not decompose because the landfill covers the polymers. Plastic is not biodegradable, but *photodegradable*. And in reality, most plastic does not ever disappear but becomes long-lasting "plastic dust" (Fig. 14.47).

Non-biodegradable waste is a type of waste that cannot be broken down into its base compounds by microorganisms, air, moisture, or soil in a reasonable amount of time (Fig. 14.48).

Plastics sustain the petroleum industry, an industry with deep pockets. This industry can have reaches into our government and our political leaders and can

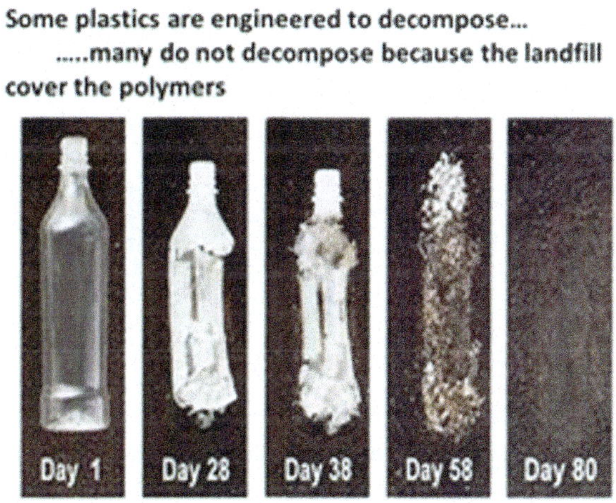

Fig. 14.46 Decomposition of plastic

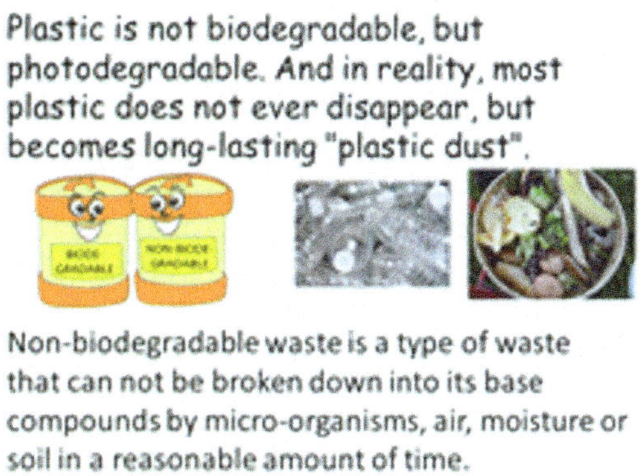

Fig. 14.47 Photodegradable

Fig. 14.48 Oil industry and plastic

Fig. 14.49 Sustainability

control legislative outcomes which would circumvent the problems that surround polymers (Fig. 14.49).

Many areas of the country have petroleum sustaining the livelihood of communities. There is little concern for the ecological effects that the industry laden on our environment (Fig. 14.50).

The Environmental Impact Assessment/*Environmental assessment* (*EA*) is the assessment of the environmental consequences (positive and negative) of a plan, policy, program, or actual projects prior to the decision to move forward with the proposed action.

Environmental assessments may be governed by rules of administrative procedure regarding public participation and documentation of decision-making and may be subject to judicial review. Ideally speaking the procedures could allow for establishing procedures that could best serve the community, or it could be a bureaucratic mechanism to prevent any happening that would go contrary of the industry. In this case the agency may be a hollow bit of lost money (Fig. 14.51).

One fact that is known is that plastic is an endangerment to wild life (Fig. 14.52).

Fig. 14.50 Environmental assessment (EA)

Fig. 14.51 Danger to wildlife

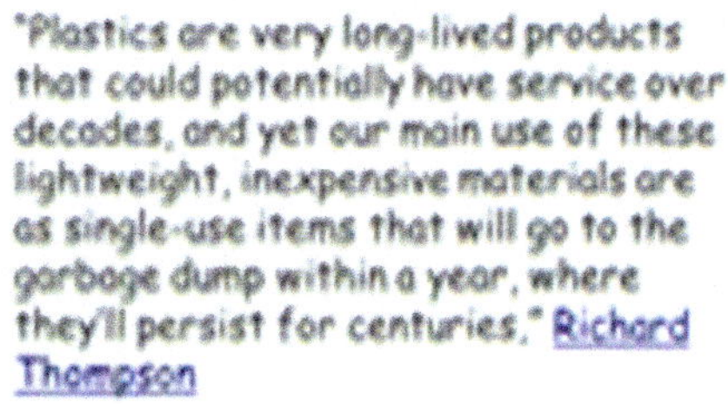

"Plastics are very long-lived products that could potentially have service over decades, and yet our main use of these lightweight, inexpensive materials are as single-use items that will go to the garbage dump within a year, where they'll persist for centuries," Richard Thompson

Fig. 14.52 Richard Thompson

> Plastics are very long-lived products that could potentially have service over decades, and yet our main use of these lightweight, inexpensive materials are as single-use items that will go to the garbage dump within a year, where they'll persist for centuries. (Richard Thompson)

Have we oversaturated the everyday environment with polymers? (Fig. 14.53) The production of plastic materials requires large quantities of crude oil as the main ingredient. Increasing oil prices and difficulties in obtaining oil have meant that the costs of producing plastics are gradually rising. The process of producing plastics also requires large amounts of energy and increases the strain on the environment to provide raw materials for a wasteful material that is highly difficult to recycle. The process of plastic production produces large quantities of toxic emissions, which contribute to the greenhouse gas effect. Polymers can be burned or incinerated, releasing a great amount of heat energy which could be used to heat homes or create electricity. However, this heat is a result of the burning of polymers, producing excessive amounts of carbon dioxide, which can aid in global warming. Toxic gases are produced unless the incineration is done at high temperatures (Figs. 14.54 and 14.55).

What problems are there in our use of plastics? It is the disposal of the material. In the disposition of the material, the effects on the environment are large. Or are they?

Burning plastic, as we know it, creates extremely polluting by-products, as evidenced by the black smoke produced. Instead of directly setting the cup aflame with a match in the open air, the team's reactor heats the material to a whopping 800 °C in a completely oxygen-free environment. This causes the plastic to become a gas, which is then mixed with air before it is burned as a clean fuel (Fig. 14.56).

The patent-pending process gave way to another surprising result: when the researchers introduced a bit of stainless steel into the reactor, they found it acted as a catalyst for growing uniform carbon nanotubes. Thus, not only can our plastic waste problem be rerouted to generate electricity, it can also be upcycled to generate

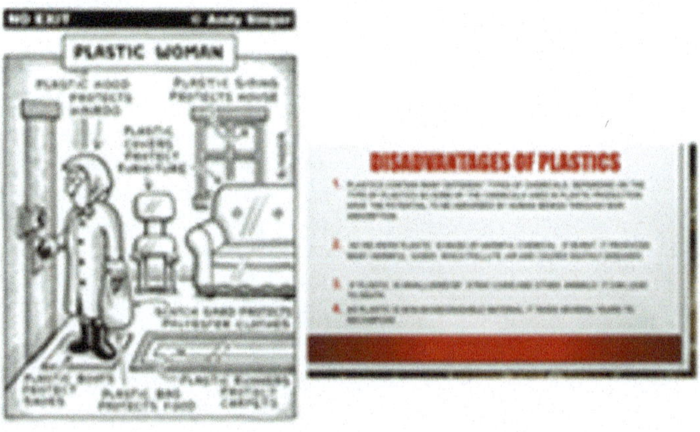

Fig. 14.53 Disadvantages of plastics

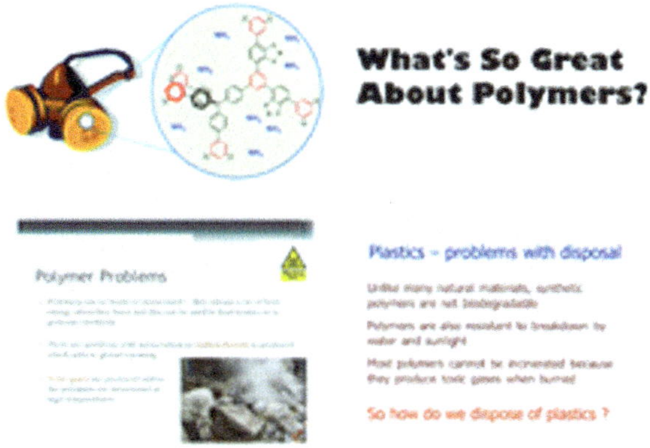

Fig. 14.54 Slide 57

Fig. 14.55 Slide 58

one of the market's most popular new materials, burning plastic as cleanly as natural gas (Fig. 14.57).

Between recycling, reusing, have a new heat source, and the creation of electrical power, polymer usage, may have changed the stereotype of a pollutant material which endangers our environment, to a material that is user friendly.

With the amount of plastics introduced into our society, the stopping of its use doesn't seem feasible (Fig. 14.58).

Today, plastics frequently replace most traditional material such as wood, metal, glass, leather, paper, and rubber because they can be lighter, stronger, corrosion-resistant, very durable, and a better insulator. These properties often make plastics more economical to use than other materials. They are used throughout industry and business because of the ease of their use; they influence the way we dress, the way we enjoy ourselves, and the way we live. Many plastics have become household names – nylon, polyester, and so on (Fig. 14.59).

Fig. 14.56 Slide 59. December 5, 2013 by Angela Herring, Northeastern University, Professor Yiannis Levendis and doctoral candidate Chuanwei Zhou with their reactor. (Credit: Brooks Canaday)

Fig. 14.57 Slide 60

Fig. 14.58 Slide 61

Traditional Materials

Stones, bricks, cement, lime and timber are the traditional materials used for civil engineering construction
(http://www.newagepublishers.com/samplechapter/001798.pdf)

Non Traditional Materials

Non-traditional primary sources can be found in a diverse number of formats including audiovisual materials, artifacts, photographs, digital files, or other formats. These non-traditional materials can provide context and enrichment for the manuscripts, correspondence, books, and other printed matter they accompany in an archival collection. Collections of non-traditional materials can stand on their own as well, evidenced by the growing collections of media, photographic, and digital archives.

Fig. 14.59 Slide 62

Fig. 14.60 Slide 63

There have been new approaches in the discard of polymer waste. Is there any other? If the issue of plastic disposal is removed from the table, then what is the argument in the use of polymers? (Fig. 14.60).

Is it litter? Is litter a problem attributed to polymers, or is it an issue that is cause by mankind? (Fig. 14.61).

Changes will need to occur, especially in the collection of polymer waste. It cannot be simply thrown in a landfill. And the reactors that would convert the polymers to clean reusable material need to be built. Are the people going to complain "not in my backyard?" or are they going to recognize the new source of fuel that would be available to create electricity?

Times are changing! (Fig. 14.62)

What are the pros and cons of using polymers? Cost is the major pro and disposal is the major con. The major pro will not be eliminated, and the major con is in the process of elimination. A new product usually creates new problems. These problems can be resolved, but these problems cannot be done instantaneously. It takes

Fig. 14.61 Slide 64

Fig. 14.62 Slide 65

What are the Pros and Cons of using Polymers?

Pro	Con
COST	Disposal

Fig. 14.63 Slide 66

Educated....

......Jury

time and research. It is easy to complain and ask for immediate answers; however, solutions are not always readily available. The government cannot solve all problems; at times it takes science to find the cure to what ails us (Fig. 14.63).

As science teachers we have to be objective, open minded, and a source of preparation for the youth that have been placed into our charge. We should reframe from bias, and this is the most difficult and presents material on a factual, level, manner. What are the pros and cons to the use of polymers? I will leave that up to you to decide; I can present the facts, void of opinion, but the final decision is up to the jury of peers. You the educated

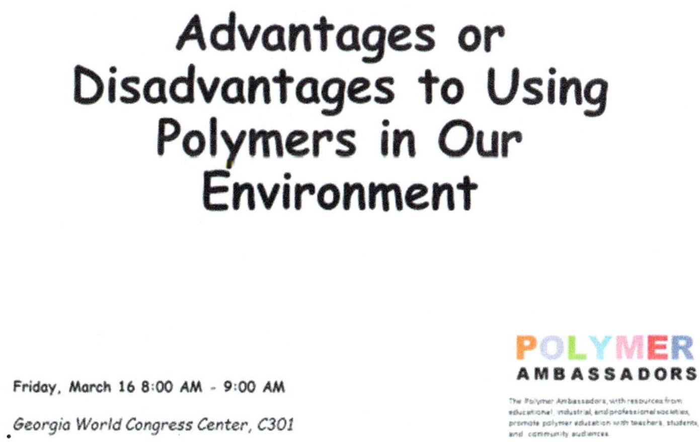

Chapter 15
Polymer Identification Code

When polymers are referenced, the basic understanding is limited. All polymers are not created equal; there are differences; and it is important to be able to identify the differences in the world of polymers. The chapter is dedicated to testing polymers to determine the type of polymer it is and understanding how recycling is limited to the particular types of polymer.

A designated polymer identification code is established by the Society of the Plastic Industry (SPI) to identify the polymer product containing a set of specific properties. The purpose of this code is to distinguish plastic sufficiently for effectual recycling. The code is represented by an arrow in a clockwise rotation, forming a triangular shape, enclosing a number. The number assigned *to the polymer product identifies a particular group of plastics possessing a particular set of properties.* The number not used indicates how difficult the item is to recycle or how often the plastic has been recycled. This code was established in 1988 and is internationally recognized and accepted.

- Polyethylene terephthalate (PETE) – the material used to produce most plastic bottles for fizzy drinks, ovenproof food trays, roasting bags, and audio and video tapes. Code 1 (Fig. 15.1):
- High-density polyethylene (HDPE) – this product is used in the creation of plastic bottles for milk, fruit juices, household cleaners, and chemicals, motor oil containers, some carrier bags, and most aerosol caps. Code 2 (Fig. 15.2):
- Polyvinyl chloride (PVC) – this product is used in the creation of plastic bottles for mineral water, fruit squash, cooking oil, and shampoo, sandwich packs, cake packs, food packaging trays, DIY blister packs, baby care product containers, cling film, ring binder cover, records and watch straps. Code 3 (Fig. 15.3):

R. A. Wesolowski et al., *The World of Materials*,
https://doi.org/10.1007/978-3-030-17847-5_15

Fig. 15.1 A designated polymer identification code is established by the Society of the Plastic Industry (SPI) Code 1

Fig. 15.2 A designated polymer identification code is established by the Society of the Plastic Industry (SPI) Code 2

Fig. 15.3 A designated polymer identification code is established by the Society of the Plastic Industry (SPI) Code 3

Fig. 15.4 A designated polymer identification code is established by the Society of the Plastic Industry (SPI) Code 4

- Low-density polyethylene (LDPE) – this product is used in the creation of a "Jif" lemon juice container, some squeeze containers for sauces and cosmetics, plastic films (shrink wrap), sacks, freezer bags, carrier bags that are not crinkly, disposable pipettes, some aerosol caps, some plant pots, and ink tubes of ballpoint pens. Code 4 (Fig. 15.4):
- Polypropylene (PP) – this product is used in the creation of plastic straws, containers for soft cheeses and fats, some margarine tubs, microwaveable food tubs and trays, film snack bags, ketchup bottles, and bottle caps. Code 5 (Fig. 15.5):
- Polystyrene (PS) – this product is used in the creation of yogurt tubs, margarine tubs, clear egg boxes, food packaging trays, plastic cutlery, cups, clear plastic glasses, ballpoint pen cases, cassette boxes, and plastic coat hangers. Code 6 (Fig. 15.6):

Fig. 15.5 A designated polymer identification code is established by the Society of the Plastic Industry (SPI) Code 5

Fig. 15.6 A designated polymer identification code is established by the Society of the Plastic Industry (SPI) Code 6

Fig. 15.7 A designated polymer identification code is established by the Society of the Plastic Industry (SPI) Code 7

- Expanded polystyrene (EPS) – this product is used in the creation of fast-food packaging, meat packaging trays, and egg boxes. No code necessary (Fig. 15.7).
- Other plastics such as acrylic, nylon, polycarbonate, and polylactic acid (a bio-plastic) and multilayer combinations of various plastics, bottles, plastic lumber applications, headlight lenses, and safety shields/glasses use the same code as expanded polystyrene.

The use of all of the aforementioned symbols may be irrelevant being that most municipalities recycle only PETE (polyethylene terephthalate) or HDPE (high-density polyethylene, HDPE), code numbers 1 and 2.

The selection of polymers for particular uses is designated by its functional requirement and costs. With the advancements in polymer research, the varieties of polymers are ever-changing. This results in the codes needing to be adjusted for future expanses of polymers.

In the SPI system, the original number sequence for polymers was one to six. Addition to the system must be undertaken on an international agreed upon basis, keeping with the objectives in maintaining a universally recognized approach ("Report on polymer").

Polymer Identification Lab

Objective: The purpose of this lab is to identify the types of polymers presented.

Materials:

Container with six polymer samples
Isopropyl alcohol solution 45.5% by volume (45.5 ml isopropyl alcohol to 100 ml water or diluting 65 ml 70% isopropyl alcohol to 100 ml water)
Mazola corn oil
Copper wire
Cork to fit 18 mm test tubes
Glass stirring rods
Acetone
Bunsen burner
Beaker, 50 ml
Beaker 250 m
Tongs or forceps
Ring stand with gauze

Test Procedure

Water Test: Place about 5 ml of water in a test tube. Start with one of the pieces of plastic. Note whether the samples floats or sinks. If the samples do not respond in the same manner, try another sample. Save the samples that floated for the alcohol test.

Isopropyl Alcohol Test: Place 5 ml of isopropyl alcohol in a test tube. Using one of the samples that floated, place the sample in the test tube, and using the glass stirring rod, attempt to make it sink.

Oil Test: Place 5 ml of Mazola oil into a test tube. Using one of the samples that floated in the isopropyl alcohol, place it into the oil test tube. Use the stirring rod to remove bubbles and attempt to make the samples sink. Test all samples determining whether they sink or float (Table 15.1).

Copper Wire Test: This test uses the samples *that sank in the water*. Use a piece of copper wire about 5-cm-long, and place the other end into a cork (to be used as a handle). Place the copper wire into a flame until it is red hot and the flame has no green color. Place the copper wire on the plastic you are testing. A small piece of plastic may melt onto the copper wire. Place the wire with a small sample of plastic into the flame. You should see a *flash* of luminous flame (a yellow-orange color). If the color is green, the sample contains chlorine.

Table 15.1 Density chart of materials in the designated polymer identification code is established by the Society of the Plastic Industry (SPI)

Substance	Density in g/ml
Water	1
PETE	1.38–1.39
HDPE	0.95–0.97
PVC	1.16–1.35
LDP	0.92–0.94
PP	0.90–0.91
PS	1.05–1.07

Acetone Test: Place about 10 ml of acetone in a 50 ml beaker. Use samples that did not give a green color flame. Place the plastic into the acetone for about 20 seconds. Remove the sample and place it between your fingers. A positive reaction occurs if the sample is *soft and sticky*. If it reacts then dispose of it in the trash; the test is completed.

Heat Test: Place about 100 ml in a 250 ml beaker and heat to boiling. Use the samples that did not pass the acetone test. Place the sample into the water using tongs for about 30 seconds. Place the sample between your fingers to determine if it softens. Positive reaction is a soft sampling. If the sample is soft, it is a positive reaction and can be discarded (Fig. 15.8).

- Polyethylene terephthalate (PETE) – the material used to produce most plastic bottles for fizzy drinks, ovenproof food trays, roasting bags, and audio and video tapes. Code 1:

- High-density polyethylene (HDPE) – this product is used in the creation of plastic bottles for milk, fruit juices, household cleaners, and chemicals, motor oil containers, some carrier bags, and most aerosol caps. Code 2:

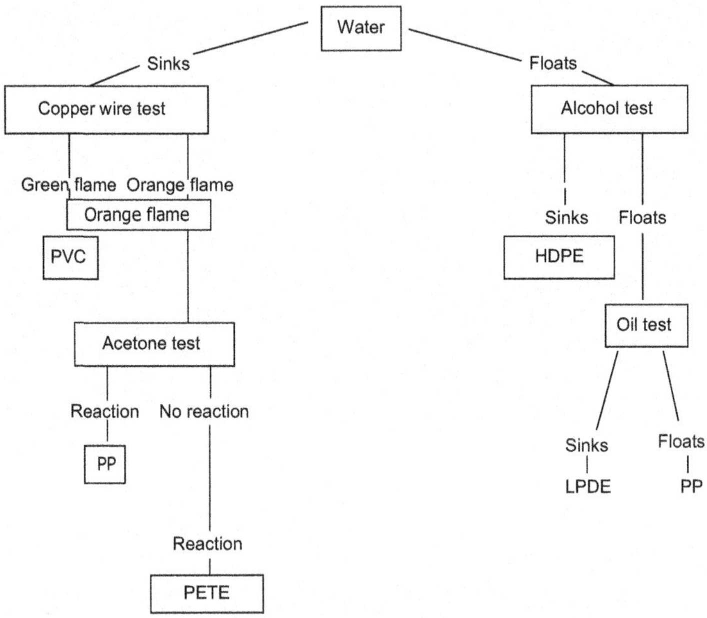

Fig. 15.8 Flowchart, testing polymer samples

- Polyvinyl chloride (PVC) – this product is used in the creation of plastic bottles for mineral water, fruit squash, cooking oil, and shampoo, sandwich packs, cake packs, food packaging trays, DIY blister packs, baby care product containers, cling film, ring binder cover, records and watch straps. Code 3:

- Low-density polyethylene (LDPE) – this product is used in the creation of a "Jif" lemon juice container, some squeeze containers for sauces and cosmetics, plastic films (shrink wrap), sacks, freezer bags, carrier bags that are not crinkly, disposable pipettes, some aerosol caps, some plant pots, and ink tubes of ballpoint pens. Code 4:

- Polypropylene (PP) – this product is used in the creation of plastic straws, containers for soft cheeses and fats, some margarine tubs, microwaveable food tubs and trays, film snack bags, ketchup bottles, and bottle caps. Code 5:

- Polystyrene (PS) – this product is used in the creation of yogurt tubs, margarine tubs, clear egg boxes, food packaging trays, plastic cutlery, cups, clear plastic glasses, ballpoint pen cases, cassette boxes, and plastic coat hangers. Code 6:

- Expanded polystyrene (EPS) – this product is used in the creation of fast-food packaging, meat packaging trays, and egg boxes. No code necessary.

- Other plastics such as acrylic, nylon, polycarbonate, and polylactic acid (a bioplastic) and multilayer combinations of various plastics, bottles, plastic lumber applications, headlight lenses, and safety shields/glasses use the same code as expanded polystyrene.

Chapter 16
Composites

Combining materials can make a new product. When the product made creates a better product, unique to itself, then a composite is created. The uniqueness of the product is a combination of two or more materials, and the combination is better than either of the initial materials. Consider a piece of Styrofoam; if it is bent, it will break. Tape a piece of duct tape on the Styrofoam. Now bend the combination of materials and what will happen? Welcome to composites.

Composites

A composite material or composite for short is a material made from two or more specific materials unique to themselves, that is, having specific physical or chemical properties, when combined, create a new material that have characteristics different from either of the initially used properties. The physical or chemical properties referred to are categorized as "matrix" or "reinforcement". The "reinforcement" can be viewed as the property that gives strength to the composite when under stress. The "matrix" transfers the stress to this "reinforcement." By investigating composites, a better understanding of the matrix and reinforcements will be accomplished.

Combining materials and maintaining the uniqueness, at the macroscopic or microscopic scale, are a composite. The combining of materials creates a final product with greater characteristics than either of the two products initially possesses. This is also known as a synergistic effect. Glass fiber-reinforced plastics (GRP) today provide an alternative in its application to conventional structural materials such as concrete, steel, aluminum, and wood. "In particular it combines high strength and low weight, while at the same time it is non-corrosive and has thermal and electrical insulation properties. It can also be machined like wood using diamond-tool equipment" ("Why use composites?").

A composite material is composed of "two or more distinct materials (such as base, binder, filler, reinforcer) that do not merge (do not lose their individual identities) and a matrix material such as a polymer resin or concrete but still impart their properties to the product resulting from their combination" ("Composite materials").

R. A. Wesolowski et al., *The World of Materials*,
https://doi.org/10.1007/978-3-030-17847-5_16

Fig. 16.1 Materials property chart on composites

Strength	Thermal insulation
Stiffness	Thermal conductivity
Corrosion resistance	Accoustical insulation
Weight	Temperature dependence
Appearance	Wear resistance

Material properties that can be improved upon by forming a composite are (Fig. 16.1):

A composite is not unique to man-made materials. There are many natural composite materials. Wood is a composite, where the lignin matrix is reinforced with cellulose fiber. Bones are composites having the bone plates reinforced with collagen. Seashells are viewed as composites as the calcium carbonate is held together by proteins.

Man used composites when they mixed mud with straw making a brick. This material was initially discovered in the area known as Israel. Materials found in Egypt, dated back to 1500 BC, showed a lamination of wood forming plywood. Also similar to the application of straw and clay as found in Israel, the same straw clay mixture was used in Egypt to reinforce walls. Around 1800, layering of metal was used to form swords and armor. During the 1950–1960, materials of light weight and rigidity and high strength were needed for the aeronautic industry. During the "Cold War," advances were made in the space race between the United States and the USSR. Glass fibers are a brittle material. Surrounding the glass fiber with a polymer creates a fiberglass material which possesses more strength and reduces the brittleness of the glass fiber. The plastic material also adds to the protection of the glass fibers from outside forces. An ordinary glass plate stresses and fractures with as little as several thousand pounds per square inch of pressure; however, adding a fibrous composite the strength of the glass plate will increase between 400,000 and 1 million pounds per square inch.

The fiberglass material evolved as a result of experimentation with the use of carbon fiber. The carbon fiber was much stronger than fiberglass but more expensive to produce than fiberglass. This carbon fiber became the basic material for the military and the aerospace industry.

Forty years ago, a new product was introduced into our everyday lives, carbon fibers. There are very thin materials approximately a tenth of the thickness of human hair. The fibers can be bundled, woven, and made into tubes and sheets that are available for construction as well as supplied in the textile industry for molding. This material, carbon fibers, has gone to space and back. Used in the space industry and aeronautics industry, carbon fiber's weight to strength ratio far exceeds that of any metal. The material reduces the weight of the product, thereby reducing the energy needed to move the product.

The use of carbon fiber to strengthen damaged cruciate ligaments in the knee is being researched, but probably the most well-known medical use is that of prosthetics – artificial limbs. South African athlete Oscar Pistorius brought carbon fiber limbs to prominence when the International Association of Athletics Federations failed to ban him from competing in the Beijing Olympics. His controversial carbon fiber right leg was said to give him an unfair advantage, and there is still considerable debate about this. (Johnson, T)

The two parts of the composite are the binding material and the matrix material. The binding material is the substance that encompasses the other material. It is the cement that holds the sand and gravel together. The matrix material is the glass fiber, the ceramic, the metal, and carbon elements that are bonded together. The joining of the binding material and the matrix materials makes the composite, and the composite is stronger than either of the combined materials alone. Basic types of composites fall under three categories, laminate composites, fiber-reinforced composites, and particle-reinforced composites. The laminated composite is a layered material. Consider a piece of plywood. The individual sheets of plywood are glued together. The sheets are staggered, so the grain of the wood from one sheet to another is 90° to each other. This layering reinforces each particular sheet making the whole of the composite stronger than any one sheet (Fig. 16.2).

Fiber-reinforced composite is made of fibers which are held together by a matrix material such as glue. The fiber transfers its stress from one fiber to another distributing the load across the longitudinal direction of the fiber. The matrix binder holds the fibers together and transfers the stress between fibers. Materials such as epoxy, polyester, and polyurethane are examples of a matrix binder. Aluminum, copper, and lead are examples of a fiber material. An example of the fiber-reinforced composite would be particle board. Particle-reinforced composites are materials made of particles that are reinforced by a matrix binder. The particle-reinforced composite is the cheapest composite to be produced. The composite is rated by the size of the particles that are being held together. The smaller particles used, the stronger the composite. Concrete has large particle size that is used to bond itself with sand and gravel. This process has been known in the time of early Egypt, 400 BC (Clay and Pottery 2015) (Figs. 16.3 and 16.4).

Regardless of the type of composite manufactured, it is imperative to minimize the chance of failure. In any type of particle-reinforced composite, there exists the possibility of the creation of a void between to fiber and the matrix. This may be a result of a simple air bubble being introduced while the composite is setting up, i.e.,

Fig. 16.2 Plywood, Wikipedia, the free encyclopedia, en.wikipedia.org, 2288 × 1712, Search by image Average-quality plywood with show veneer

Fig. 16.3 Particle board, Wikipedia, the free encyclopedia, en. wikipedia.org 20, × 199, Search by image cross section of particle board

Fig. 16.4 Concrete, patocean.com, 651 × 300, Search by image

solidifying. Another type of defect that may occur is when the laminated material may peel away are a result of a breakdown of binding material to the laminate. This may occur if an improper application of a binding agent to the laminate or upon the application of the binding agent an inconsistent pressure is applied to the laminated materials during the curing ultimately reducing the strength of the material. The result of the defect may range from minimal when a laminated material has a surface separating and the surface may be repaired to a catastrophic collapse where a defect in a material distributes its load to other composites which may not be capable of sustaining the payload resulting in a total failure.

What advantage is there to using a composite material? The first major importance found in the composite is its weight. When compared to metals and woods, composites are extremely lightweight. When attempting to reduce fuel consumption in the automobile industry or the aeronautic industry, weight becomes a major concern (Fig. 16.5).

Reduction in the weight of a car or a plane would increase the efficiency of the vehicle in the distance traveled to the fuel expended. The reduction in the weight of a car will have an effect on the speed of the vehicle. This increase in the speed is without any change in the engine dynamics. The same thrust on a lighter object would render a greater speed. This would decrease travel time without any additional expense in fuel consumption. The composite may increase the strength of the product. A metal has strength in all directions, but a composite may be engineered to increase its strength in a particular direction. Additionally, the strength of a composite compared to its weight is significantly better. To increase strength in a metal

Fig. 16.5 Natural fibers for automobiles, www.naturalfibersforautomobiles.com, 640 × 480, Search by Image

requires an increase in the material which immediately increases the weight. Composites are strong without being heavy.

Composites are used when the product is subject to environmental conditions. Corrosion may be prevented. It there is a chemical concern, the composite resists damage by the chemical.

The structure of a composite may be designed into complicated shapes that are not possible using metals and woods. Composites can be designed to be radar transparent. The stealth bomber is designed using composite material enabling the aircraft to become free of radar detection. The most significant advantage of the composite is the material has a long life requiring little maintenance (Why Composites? 2015).

References

Clay and Pottery – Brief History of Pottery and Ceramic. (n.d.). Retrieved March 18, 2015, from http://lakesidepottery.com/HTML Text/brief history of clay_pottery.htm

Why Composites? (2015, January 1). Retrieved March 17, 2015, from http://www.prcmix.com/why-composites/adv-composites.php

Chapter 17
Making and Testing Composites

Theory is great, but theory does not beat getting your hands dirty. Tests can be conducted determining the strength of the product made. A "three-point test" is a simple test where the product is permitted to reach its maximum stress point. Investigating the deflection point is another test used in industry to measure flexibility. Creating the structure of the material, its appearance, so that it can be tested is also critical. To use the three-point test, a beam must be created. It may be important to create a cylinder to use a compression test. Here is where making and testing composites becomes enjoyable.

There are various ways to illustrate how composites are stronger materials than the components parts of the composite. Using a piece of Styrofoam, approximately 12″ × 1″ × 1/2″, have a volunteer place the material over their head, and begin to bend the material ever so slowly. The Styrofoam material will eventually break, showing little ability of restraining the force being applied. Take a send piece of Styrofoam of similar dimensions. Apply a strip of *duct* on one side of Styrofoam thereby making a composite material. With the "*duct*" tape on the top, have the volunteer repeat the bending procedure. The composite Styrofoam will bend but not break. The composite material has a new and different characteristic; it can withstand the force applied and is therefore stronger than the original material. What happens when the composite Styrofoam has the *duct* tape on the bottom? When the volunteer places the composite Styrofoam over their head and applies a force as done in the past, will the material created withstand the force or not? What would occur if a new piece of Styrofoam of similar dimensions has a piece of *duct* tape wrapped entirely around the Styrofoam and a similar force is applied as in the previous two cases? The tape and Styrofoam system is a form of lamination. "*Lamination* is the technique of manufacturing a material in multiple layers, so that the composite material achieves improved strength, stability, sound insulation, appearance or other properties from the use of differing materials. A laminate is usually permanently assembled by heat, pressure, welding, or adhesives" (Lamination 2015).

R. A. Wesolowski et al., *The World of Materials*,
https://doi.org/10.1007/978-3-030-17847-5_17

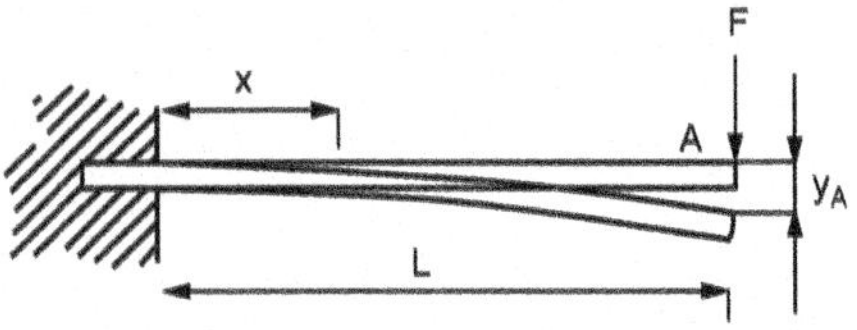

Fig. 17.1 Deflection structures, www.codeclogs.com, 537 × 158 Search by Image

A second type of an example to demonstrate the properties of a composite is seen in the usage of a wooden stick. Fasten a wooden stick to a lab table, and deflect or bend the stick by applying masses or weights to the stick as illustrated below (Fig. 17.1).

Recording the data of mass applied to deflection, y_A as various masses are applied, allows for a graph to be created illustrating the degrees of deflection. This process may be continued until the stick is broken. Gather two additional stick and glue the sticks together. The gluing of the sticks will need to be accomplished in advance in order to have a testing session performed comparing the single stick with the glued stick. Repeat the process of fastening the glued sticks to a lab table in order to apply masses, and record the deflection. Compare the findings of the two sticks glued together to the single stick. Compare the graphs of mass versus deflection. What conclusion can be made regarding the composite structure to the single stick if a student believes that there should be a two to one ratio of the deflections because there are two sticks compared to one? Repeat the deflection procedure of the single stick with simply having two sticks fastened to the lab table but not glued together, and collect data for mass versus deflection. The comparison of this sample, the two sticks, to the glued sticks, is a comparison of a composite to a non-composite material. The gluing is significant and should be seen in the comparison of the graphs. This activity can and should open the discussion of the significance of a composite material.

Creating a concrete hockey puck or a concrete beam is an interesting activity in showing the strength of composites. Have students make a hockey puck. One hockey puck should be a test sample and be simply the concrete placed in a mold. The mold can be a plastic dish such as a petri dish or a plastic dish found at a discount store. Apply the same amount of concrete in each student's dish, allowing for the same amount of water to be added to the mixture. Before the mixture is complete, have the students add an aggregate to the concrete water mixture. The aggregate may be straw, fiber, paper clips, or whatever the students wish to include. The aggregate serves as a source to strengthen the concrete mixture. Allow 3–4 days for the mixture to dry or cure.

One technique used to test the concrete strength is dropping a ball bearing onto the concrete puck. This procedure is a compression test, a test used in industry to verify the strength of the concrete mixture. Save the tubes from wrapping paper. These tubes can be used as the controlled path for the falling ball bearing onto the concrete hockey pucks. Mass out the ball bearing! Various sized ball bearing can and should be used. Also, the tubes should be of various lengths. Have a small length tube and small ball bearing be used to test all the samples and compare the indentation. A second round of impacts may be performed using the same lengthened tube with a larger ball bearing. A comparison discussion can follow looking at what the aggregate was and how was it applied. This process of testing can expand to using a longer tube with the same masses. This procedure would continue until the concrete hockey pucks are broken. Then a rating can be made as to which procedure created the strongest puck.

Making a beam from concrete can also allow a test on strength. Take Styrofoam and make a mold for a beam, ~2″ × 2″ × 2″ (Fig. 17.2).

In a manner similar to the hockey pucks, a sample will be created with pure concrete. All other samples will have an aggregate added to the concrete. Allow 3–4 days drying time.

To test the beam, apply a three-point test. Support the beam over two beams; these beams can be the backs of a chair or the supports of a ring stand (Fig. 17.3).

The masses would be added until the beam is broken.

The activity used is dependent on the supplies of the class. Nevertheless, the activities are enjoyed by the participants and are educational. The level of discussion is limited to the level of the class.

Upon completion of the composite test, students may be permitted to make a PowerPoint presentation of the creation of the beam or puck and the results of the test. Pictures can be taken as the test is applied. The pictures can be posted to demonstrate the activities the students are engaged.

Fig. 17.2 Styrofoam mold

Fig. 17.3 Three-point test

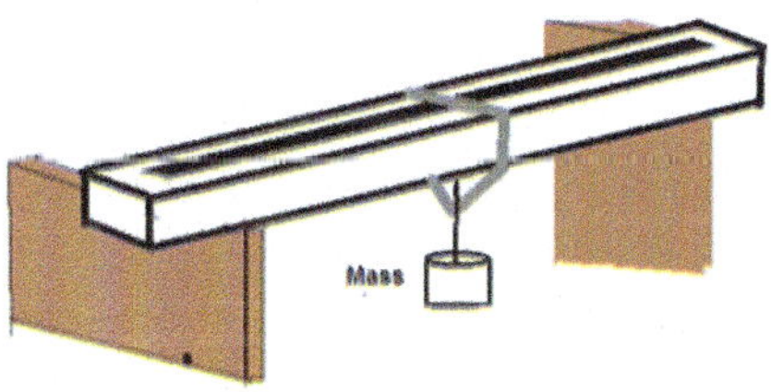

Reference

Lamination. (2015). Wikipedia, Carter, Richard. *Method of laminating multiple layers*. Google Patents. Retrieved 15 February 2015.

Quizzes per Chapter

Upon Completing the Reading of Chapter 1, Answer the Questions

Chapter 1

Introduction to Materials

Name ______________________

1. **What considerations are made in the selection of materials?**

   ____

   ____

   ____

   ____

2. **Identify five materials and products that are on the cutting edge for society.**
   1. ______________________________________
   2. ______________________________________
   3. ______________________________________
   4. ______________________________________
   5. ______________________________________

3. **What are the five steps to scientific method?**
   1. ______________________________________
   2. ______________________________________
   3. ______________________________________
   4. ______________________________________
   5. ______________________________________

4. **What is a non-Newtonian fluid?**

   ____

   ____

   ____

   ____

5. **Define dilatant and thixotropic materials**

   ____

   ____

   ____

   ____

R. A. Wesolowski et al., *The World of Materials*,
https://doi.org/10.1007/978-3-030-17847-5

Upon Completing the Reading of Chapter 2, Answer the Questions

Chapter 2

Classification of Matter

Name ______________________

1. **What is the difference between science and technology?**

____
____
____
____

2. **What does material generally consists of?**

____
____
____
____

3. **What is the difference between elements, compounds, and mixtures?**

____
____
____
____

4. **What is the difference between crystalline and amorphous structures?**

____
____
____
____

5. To understand the composition of material, an awareness of the atomic structure of the material is essential. Consider the structure of the atom. It is to be recognized as the smallest particle consisting of the properties of a particular element. It contains three basic subatomic particles: ______________, ______________, and ______________.

Upon Completing the Reading of Chapter 3, Answer the Questions

Chapter 3

Atomic Structure and Periodic Table

Name ____________________________

1. **What are cations and anions?**

____
____
____
____

2. **On the periodic table, the rows are referred to as __________ and the columns are referred to as _________**

3. **Which elements had complete outer shells? Give the names and symbols of each.**

____
____
____
____

4. **What do you notice about the locations of the elements from question 3?**

____
____
____
____

5. **What do you notice about the number of energy levels or shells as you go down a group or column in the periodic table (H - Li - Na)?**

____
____
____
____

Upon Completing the Reading of Chapter 4, Answer the Questions

Chapter 4

Crystal Properties

Name ____________________________

1. **A ___________ is the alignment of atoms that defines the material.**

2. **Define defect.**

   ____

   ____

   ____

3. **If the impurity is placed in an atom's void, the defect is referred to as**

   ____

4. **What is a grain boundary?**

   ____

   ____

   ____

5. **A solid material made of organized parts or constituents (i.e., atoms, molecules, or ions) that are microscopically ordered in all directions are called**

   ____

Upon Completing the Reading of Chapter 5, Answer the Questions

Chapter 5

Crystalline and Amorphous Solids

Name ______________________

1. **Why we study crystal structure?**

____
____
____
____
____

2. **When does iron transition from ferromagnetic to paramagnetic?**

____
____
____
____

3. **What is the nucleation point?**

____
____
____

4. **An example of crystallization resulting from a chemical reaction would be a** ______________

5. **What is an allotrope?**

____
____
____
____
____

Upon Completing the Reading of Chapter 6, Answer the Questions

Chapter 6

Mechanical Properties

Name ______________________________

1. What is the difference between a crystalline versus an amorphous structure of metals?

   ____

   ____

2. An amorphous material, glass, is created due to a __________ prior to crystallization. Glass becomes ________ and often optically __________

3. A ____________ is referred to as a discontinuity whose size, shape, orientation, or location makes it detrimental to the useful service of the part that it occurs (Reardon, 2012), which may enhance a product, strengthening it or even formulating a superior product.

4. Name three types of point defects and define them.

   1. ______________ -

      ____

      ____

      ____

   2. ______________ -

      ____

      ____

      ____

   3. ___________-

      ____

      ____

      ____

5. What is a grain boundary?

   ____

   ____

Upon Completing the Reading of Chapter 6, Answer the Questions

Chapter 6a

Mechanical Properties

Name ____________________________

1. What is the difference between elasticity and plasticity?

2. What is creep, and how does it differ from fracture?

3. What is forging?

4. Why we use forging?

5. What is "necking"?

Upon Completing the Reading of Chapter 7, Answer the Questions

Chapter 7

Semiconductors

Name ___________________________

1. What is a conductor and an insulator? How does a semiconductor compares to either of these terms?

____
____
____
____

2. What is ionization?

____
____
____
____

3. A semiconductor is not a good conductor. How can the semiconductor become a good conductor?

____
____
____
____

4. What is doping?

____
____
____

5. What does adding impurities to a semiconductor accomplish?

____
____
____
____

Upon Completing the Reading of Chapter 8, Answer the Questions

Chapter 8

Iron Wire Lab and Crystal Models

Name ___________________________

1. As heat is applied to the iron wire, the first noticeable change is:

____

____

2. At the Curie point ~ 770°C, what change occurs, and how is this change noticed?

____

____

____

3. After the wire becomes red hot, the power to the Variac is off, what occurs?

____

____

____

____

4. What is the remaining residue found on the wire?

____

____

____

____

5. What is a nucleation point?

____

____

____

____

Upon Completing the Reading of Chapter 9, Answer the Questions

Chapter 9

Graphs and Diagrams

Name ______________________________

1. **Name three pieces of information that can be obtained from a graph.**
   a. ______________________________
   b. ______________________________
   c. ______________________________

2. **What happens when heat is applied to ice?**
   ____
   ____
   ____

3. **What is the latent heat of fusion?**
   ____
   ____
   ____

4. **What is the eutectic point in a bismuth-tin phase diagram?**
   ____
   ____
   ____

Upon Completing the Reading of Chapter 10, Answer the Questions

Chapter 10

Corrosion

Name ____________________________

1. What is corrosion?

____
____
____

2. Where is corrosion found?

____
____
____

3. How does this battery work?

____
____
____
____
____

4. Name three types of corrosions.

____
____
____

5. What is a bias force?

____
____
____
____
____

Upon Completing the Reading of Chapter 11, Answer the Questions

Chapter 11

Introduction to Ceramics

Name ____________________________

1. What is a ceramic?

____

____

____

____

2. Crystal patterns of a ceramic are

a) Crystalline b) Amorphous c) Both d) Neither

3. *Kermos* is a Greek word for __________________

4. To form a ceramic, clay must be ____________ and _____________.

5. T or F: Thomas Edison invented the incandescent light bulb

6. T or F: Ceramics are good conductors of electricity but poor conductors of heat.

7. T or F: Ceramics possess a low density and a high melting point

8. T or F: All ceramics are from clay

9. T or F: Ceramics are harder and stiffer than steel

10. T or F: Ceramics are a limiting material due to its brittleness and microscopic flaws

Upon Completing the Reading of Chapter 12, Answer the Questions

Chapter 12

Biomaterials

Name ____________________________

1. **What is bioceramics?**

   ____
   ____
   ____

2. **During the beginnings of hip replacement surgery, the implant was designed to be very near the dimensions of the actual hip. It was found that the larger hip resulted in a quicker deterioration as a result of "volumetric wear." What is "volumetric wear"?**

   ____
   ____
   ____
   ____

3. **What are possible implant problems?**

   ____
   ____
   ____
   ____

4. **What is bio-inert?**

   ____
   ____
   ____
   ____

5. **What is fiber optics?**

   ____
   ____
   ____

Upon Completing the Reading of Chapter 13, Answer the Questions

Chapter 13

Polymers

Name ____________________________

1. What is a polymer?

____

____

____

____

2. Consider how spaghetti appears on a plate. This illustration resembles the arrangement of a polymer that is __________. The __________ polymer does not have a long arrangement, and this ___________ structure generally results in a transparent substance.

3. The opaque or translucent polymers are usually __________ in structure.

4. Amorphous materials are a result of rapid _________, ____________.

5. What are the two basic grouping of polymers and define each.

____

____

____

____

____

Upon Completing the Reading of Chapter 14, Answer the Questions

Chapter 14

Polymer Identification Code

NAME ______________________________

1. What is the purpose of the polymer identification code?

2. What is the purpose of the identification code number, and what does the number not indicate?

3. What is code 05 and what does PP mean?

4. Most municipalities recycle only what two types of polymers?

5. What is code 07 representing?

Upon Completing the Reading of Chapter 15, Answer the Questions

Chapter 15

Composites

NAME ______________________________

1. **Combining materials and maintaining the uniqueness, at the macroscopic or microscopic scale, is a** ______________

2. **What is a synergistic effect?**

____

____

____

____

3. **Material properties that can be improved upon by forming a composite are:**

4. **What are the two parts of a composite and define them?**

____

____

____

____

5. **What is the advantage of using a composite material?**

____

____

____

____

Additional Readings

Materials Science Resources and Web Sites

Included is a list of some useful web sites and organizations. There is some wonderful stuff here, and it is easy to spend several hours learning and enjoying yourselves without realizing where the time has gone.

http://asmcommunity.asminternational.org/portal/site/asm/

ASM International. Not just metals - has information on all materials. This association is a great resource. Hosts Materials Science teacher and student camps in the summer.

University of Washington materials science web site:

http://depts.washington.edu/mti/

Check out the following topics under "Resources":

Books and Texts.
Materials Sources.
Materials Web Links.
Teaching Resources.

http://www.polymerambassadors.org

Find out what we do and how to become a Polymer Ambassador.

http://www.chem.ox.ac.uk/icl/heyes/structure_of_solids/strucsol.html

Good site for crystals/solids with excellent links to other sites.

http://www.chemistry.co.nz/crystals_defined.htm

More basic crystal information

http://www.steel.org//AM/Template.cfm?Section=Home

American Iron and Steel Institute

R. A. Wesolowski et al., *The World of Materials*,
https://doi.org/10.1007/978-3-030-17847-5

http://www.pslc.ws/macrog/index.html
"The Macrogalleria - a Cyberwonderland of Polymer Fun". This is a great site for polymers. It is a "polymer mall" with five different levels:

Polymers are Everywhere.
Polymers Up Close and Personal.
How They Work.
Makin' Polymers.
Getting Polymers to Talk.
It would be easy to generate student research assignments from this site.

http://www.mrs.org/s_mrs/index.asp

Materials Research Society

http://www.mii.org

The Mineral Information Institute. Lots of great free stuff to order or download.

http://portal.acs.org/portal/acs/corg/content

The American Chemical Society

http://nickelalloy.com

Metal prices. Useful in research projects.

http://www.handsonplastics.com

Can order free stuff from this one!!

http://www.plasticsmuseum.org

The National Plastics Center & Museum. Order their CD entitled "Exploring the World of Plastics" if it is still available. Well worth the $10 cost.

http://www.corning.com

Lots of stuff about glass.

http://www.tms.org

Minerals, Metals, and Materials Society. Great web site and a very useful organization.

http://www.acers.org/

The American Ceramics Society

http://scifun.chem.wisc.edu/scifun.html

Lots of info – check out the "Chemical of the Week" link.

http://www.strangematterexhibit.com/

A great site with information and activities about all types of materials.

http://www.sciencemuseum.org.uk/onlinestuff.aspx

A site with great links.

http://www.thecatalyst.org/m05demos.html

Chemistry Resources For The Secondary Education Teacher On The WWW

http://www.sciencegems.com/

Frank Potter's Science Gems.

http://www.careercornerstone.org/

Sloan Career Cornerstone Center

http://www.learner.org/resources/series61.html

The World of Chemistry – movie resources

http://www.sfsa.org/tutorials/index.html

Metal casting scenarios

http://www.madehow.com

How It's Made – articles on how everyday items are made

http://manufacturing.stanford.edu/

Manufacturing Alliance – virtual tours of manufacturing processes

http://www.museumofglass.org/s02_virtual_hotshop.jsp

From the Museum of Glass in Tacoma. Neat interactive site allows you to create a 'Macchia'/spotted bowl, using glass blowing techniques

http://portal.acs.org/portal/acs/corg/content?_nfpb=true&_pageLabel=PP_TRANSITIONMAIN&node_id=1090&use_sec=false&sec_url_var=region1

ChemMatters – articles with many, many MST connections

http://www.pbs.org/wgbh/buildingbig/lab/materials.html

Interactive website by PBS. Great for exploring tension and compression strength of various materials. Incorporates real life architectural examples.

http://www.marcopolo-education.org/home.aspx

Great teacher resources – now called "Thinkfinity"

http://blog.nam.org/csbmindex.php

National Association of Manufacturers. Cool videos on how all kinds of things are made. Constantly being updated.

http://www.tipmagazine.com/tipcd.html

The Industrial Physicist – check out the past issues for all kinds of articles related to MST

http://www.ndt-ed.org/EducationResources/CommunityCollege/Materials/cc_mat_index.htm

Check out this website for great information about many different materials science topics.

http://www.thatcrystalsite.com/downloads/NaicaCrystalCave.pps#257,1,Slide1

Giant crystals in cave in Mexico

http://hazard.com/msds/
National MSDS database.

There are a large number of universities that have materials science departments as well.
One last web site that is great – you need to check it out:

http://www.howstuffworks.com

You will be amazed at the things you will learn.
Two resources for buying materials/supplies related to materials science:

IASCO (Industrial Arts Supply Co.) Educational Innovations.

1–888–919-0899 1–888–912-7171

www.IASSCO-TESCO.com, www.teachersource.com

Great supplier for metal or polymer projects. Fun, fun, fun. Coolest science “toys”.

Teacher Resources

Chapter 1

Oobleck - White Powder Lab"

Name: ______________________ Period: ________ Date: _________

Do all liquids behave the same? Answer all questions completely.

Procedure:

1. Take three cups of the selected white powder (about 1 tablespoon) and a stir stick. Record observations about the powder(s) below. Does it have an odor? Does it slip between your fingers easily?

2. Slowly add water to the powder(s) until the combination makes a putty-like material. Stir as you add water. Observe how the mixture moves as you stir with the stick. Is it difficult to move? Does it move easier as you stir it, or is it more difficult?

3. Stir rapidly and then stir slowly in the individual cups. Is there a difference in the way the mixture responds to the way you add energy? Record observations below.

R. A. Wesolowski et al., *The World of Materials*,
https://doi.org/10.1007/978-3-030-17847-5

Try putting the mixture of the individual cups of powder and water in the palm of your hand. Can you roll it into a ball? What happens if you stop trying to roll it? Record observations below.

Deposit all materials in the trash can.

1. In the space below, define the following terms and give examples of each:

Thixotropic Fluids:

Dilatant Fluids:

Chapter 2 Handout 1

Nuts and Bolts

Objective: The primary purpose of this activity is to understand the meaning of the terms, element, compound, and mixture, as well as the classification of matter.

Materials: Nine cups, washers, bolts, and hexagonal nuts (which fit the bolt).

Procedure: Place the cups in the open environment with the contents indicated for student observations.

1. Cup with six bolts.
2. Cup with six hexagonal nuts.
3. Cup with six washers.
4. Cup with three bolts and three hexagonal washers not assembled.
5. Cup with one hexagonal nut, one bolt, and one washer not assembled.
6. Cup with one hexagonal nut and bolt assembled.
7. Cup with one hexagonal nut and bolt assembled and two extra washers.
8. Cup with two hexagonal nuts, two bolts, and two washers assembled.
9. Cup with a hexagonal nut and bolt assembled and a hexagonal nut bolt and washer assembled.

Label the cups 1–9 or A–I and randomly place the cups in the environment.

Cups	Content	Element	Compound	Mixture	
1/A	Cup with six bolts			Ho	He
2/B	Cup with six hexagonal nuts				
3/C	Cup with six washers				
4/D	Cup with three bolts and three hexagonal washers not assembled				
5/E	Cup with one hexagonal nut, one bolt and one washer not assembled				
6/F	Cup with one hexagonal nut and bolt assembled				
7/G	Cup with one hexagonal nut and bolt assembled and two extra washers				
8/H	Cup with two hexagonal nuts, two bolts and two washers assembled				
9/I	Cup with a hexagonal nut and bolt assembled and a hexagonal nut bolt and washer assembled				

Under the mixture column, indicate whether the mixture is homogeneous or heterogeneous, mark appropriately.

Chapter 2 Handout 2

Consider the table of the classification of materials. Fill in the blanks.

Classification of Materials

	Metals	Ceramics/glass	Polymers
Type of matter			
Type of element			
Type of structure			
Type of bonding			

Chapter 2 Handout 3

Subatomic Particles

Fill out the table on your worksheet based on what you have learned.

Name	Electron	Proton	Neutron
Charge			
Location			
Mass			
"Job"			
Number			

Chapter 2 Handout 4

To reinforce the understanding of bonding complete the chart below:

Type of bonding	Metallic	Covalent	Ionic	Molecular forces
Types of elements used				
Givers and/or takers				
Description				
Type of material formed				
Strength of bond				
Properties produced				

Chapter 2 Quiz 1

Classification of Matter

Quiz

Name ______________________ **Date** ___________

1. Classify each of the following substances as an element, compound or mixture

Materials	Classification	Materials	Classification
Platinum		Water	
Table salt		Sulfur	
Carbon dioxide		Salad dressing	
Mayonnaise		Mercury	

2. What is the difference between homogenous and heterogeneous mixtures?

3. Which of the following is a pure substance?

   a) Water b) Milk c) Bronze d) Soda

4. In ____________, all of the atoms are the same.

   a) Compound b) Heterogeneous mixtures c) Solution
   d) Element

Chapter 3

Periodic Table

First: ***Complete the illustration on the left for all elements indicated using a computer to research the needed information that may be helpful***

Complete the indicated sections for each element by adding the element's (1) Atomic number (2) Name (3) Atomic mass (4) determine the number of protons, neutrons and Electrons (5) shade the correct circle to show what state the element is in room temperature (6) create a Bohr diagram of each element

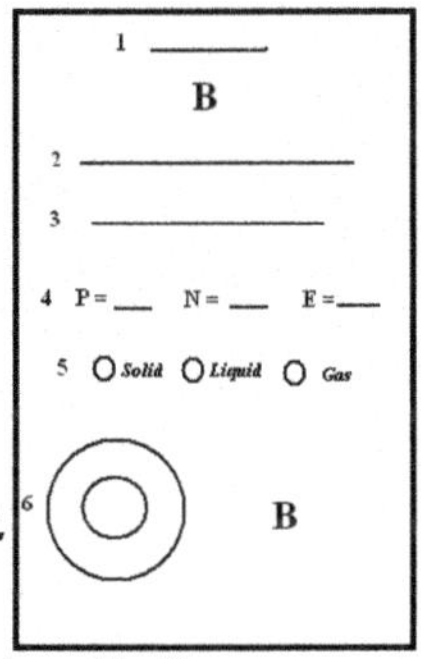

Second: ***Use color pencils to shade in the card for each element. Hydrogen is not colored***

Green: Na Li Maroon: O S Blue: Be Mg Purple: Cl F

Red: C Si Orange: B Al Yellow: He Ne Ar Beige: N P

Third: ***Cut the illustrations apart and arrange according to atomic number in the pattern shown on a large sheet of paper***

1							2
3	4	5	6	7	8	9	10
11	12	13	14	15	16	17	18

Fourth: ***After you have the cards arranges in the correct order, glue them to the paper. Add a title at the top of the paper along with your name.***

Fifth: ***Answer the given questions using the information on your periodic table.***

Questions on ***the Periodic Table***

Name ______________________________

1. Which elements had complete outer shells? Give the names and symbols of each.

_________________________ ____________________________

____

What do you notice about the locations of these elements?

2. Which elements had only one valence electron? Give the name and symbol of each.

_________________________ ____________________________

____

What do you notice about the locations of these elements?

3. What do you notice about the number of valence electrons as you move from left to right across the row or period in the periodic table? (Na → Mg → Al → Si → P → S → Cl Ar)?

4. What do you notice about the number of energy levels or shells as you go down a group or column in the periodic table (H → Li → Na)?

5. Write the name of each family at the top of the columns on your periodic table using the following information:

Alkali metals – one valence electron **Nitrogen family - five valance electrons**

Alkaline earth metals – two valence electrons **Oxygen family - six valence electrons**

Boron family – three valence electrons **Halides - seven valence electrons**

Carbon family – four valence electrons **Noble gases - complete outer shell**

6. What do you notice about the location of the elements in each family?

7. In what family would you classify hydrogen? Explain your choice!

8. In what family would each of these elements be classified?

Radium ______________________________ **Tin** ______________________________

Iodine ______________________________ **Cesium** ______________________________

9. Predict the number of valence electrons for each element based on its location in the periodic table of elements. You may need a period table.

Barium ____________ **Lead** ____________ **Bismuth** ____________
Potassium ____________

1 ________
2 ____________________
3 ____________________
4 P = ___ N = ___ E = ___
5 ○ *Solid* ○ *Liquid* ○ *Gas*
6 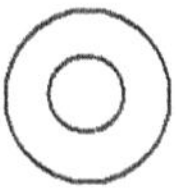

1 ________
2 ____________________
3 ____________________
4 P = ___ N = ___ E = ___
5 ○ *Solid* ○ *Liquid* ○ *Gas*
6 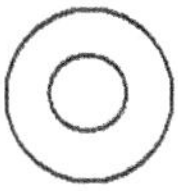

1 ________
2 ____________________
3 ____________________
4 P = ___ N = ___ E = ___
5 ○ *Solid* ○ *Liquid* ○ *Gas*
6 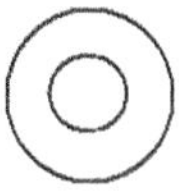

1 ________
2 ____________________
3 ____________________
4 P = ___ N = ___ E = ___
5 ○ *Solid* ○ *Liquid* ○ *Gas*
6 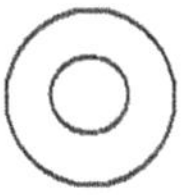

1 ________
2 ____________________
3 ____________________
4 P = ___ N = ___ E = ___
5 ○ *Solid* ○ *Liquid* ○ *Gas*
6 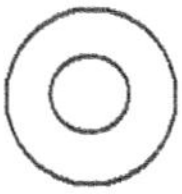

1 ________
2 ____________________
3 ____________________
4 P = ___ N = ___ E = ___
5 ○ *Solid* ○ *Liquid* ○ *Gas*
6 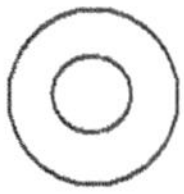

Periodic Table

First: ***Complete the illustration on the left for all elements indicated using a computer to research the needed information that may be helpful***

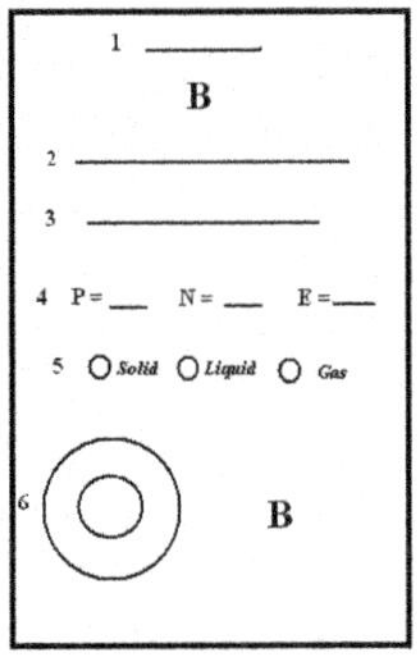

Complete the indicated sections for each element by adding the element's (1) Atomic number (2) Name (3) Atomic mass (4) determine the number of protons, neutrons and Electrons (5) shade the correct circle to show what state the element is in room temperature (6) create a Bohr diagram of each element

Second: ***Use color pencils to shade in the card for each element. Hydrogen is not colored***

Green: Na Li Maroon: O S Blue: Be Mg Purple: Cl F

Red: C Si Orange: B Al Yellow: He Ne Ar Beige: N P

Third: ***Cut the illustrations apart and arrange according to atomic number in the pattern shown on a large sheet of paper***

1							2
3	4	5	6	7	8	9	10
11	12	13	14	15	16	17	18

Fourth: ***After you have the cards arranges in the correct order, glue them to the paper. Add a title at the top of the paper along with you name***

Fifth: ***Answer the given questions using the information on your periodic table***

___1______

1 ________

2 ____________________

3 ____________________

4 P = ___ N = ___ E = ___

5 ○ Solid ○ Liquid ○ Gas

6 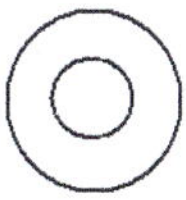

1 ________

2 ____________________

3 ____________________

4 P = ___ N = ___ E = ___

5 ○ Solid ○ Liquid ○ Gas

6 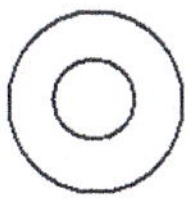

1 ________

2 ____________________

3 ____________________

4 P = ___ N = ___ E = ___

5 ○ Solid ○ Liquid ○ Gas

6 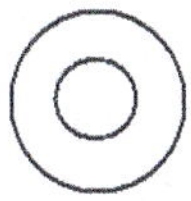

1 ________

2 ____________________

3 ____________________

4 P = ___ N = ___ E = ___

5 ○ Solid ○ Liquid ○ Gas

6 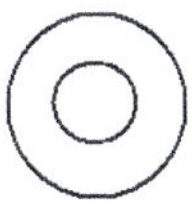

1 ________

2 ____________________

3 ____________________

4 P = ___ N = ___ E = ___

5 ○ Solid ○ Liquid ○ Gas

6 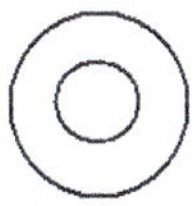

1 ________

2 ____________________

3 ____________________

4 P = ___ N = ___ E = ___

5 ○ Solid ○ Liquid ○ Gas

6 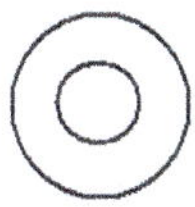

Chapter 4 Handout 1

Crystal Structures: Atomic Packing

Student Worksheet

Name_________________

1. What is a unit cell?

2. Fill in the Table Below

	Body Centered Cubic		Face Centered Cubic	
V_{sphere}				
$V_{atoms\ in\ unit\ cell}$				
$V_{unit\ cell}$				
Unit Cell: Number of atoms	**Center**	**Corner**	**Face**	**Corner**
Packing Density				

3. Platinum has an FCC crystal structure. Using the Pt atomic radius of 139 pm (1 picometer = 1×10^{-12} meters), calculate the total volume that the Pt atoms occupy in a unit cell. Also calculate the lattice constant (a) of the FCC unit cell of Platinum.

4. At 20 °C, Iron (Fe) has an atomic radius of 0.124 nm. What is the crystal structure of Iron? Calculate the lattice constant, a, for the iron unit cell.

Chapter 4 Handout 2

Across

2. ________ informs us of how many atoms that can be places in a given space or volume.
3. Replace some of the original atoms with different atoms
5. Atoms of solids are in constant ____________
6. The most common type of line defect is a _________
7. A common type of planar defect is called a grain ________

Down

1. Extra atoms inserted in the gaps between the regular atoms
2. Allotropes are also referred to as __________
4. Simplest point defect is a _________

EclipseCrossword.com

Chapter 4 Handout 3

From MAST Handbook

A Particle Model of Metals: Atomic Bb's Objective: The objective of this lab is to learn more about the basic particle model for metals.

Review of Scientific Principles: Metallic crystals are not perfect. Sometimes there are empty spaces, vacancies, where an atom should be. There are also small mismatches, dislocations, in the rows of atoms, and these are found in all metals.

Applications: Defects in the crystal structure of metals control many of their properties including hardness and ductility.

Time: 40 minutes.

Materials and Supplies: Plastic Petri dish.

Bb's.

Procedure:
Your team will be given a covered Petri dish containing Bb's. Keep the dish flat on the table, so no Bb's spill out! Answer the questions which correspond to a given procedure. Try to write complete idea answers that increases understanding.

1. The Bb's represent the atoms in a metal.
2. If you move the Petri dish back and forth, the atoms move. This simulates the movement of atoms in a metal when it is heated.
3. Move the dish back and forth and try to get the atoms arranged as neatly as possible. Slowing the motion of the dish and gradually stopping it simulates the formation of the crystal.
4. Make a sketch below showing how the Bb's are arranged. You don't have to draw all the Bb's!
5. Make a sketch below showing the arrangement of Bb's around an empty space. When this happens in the metal, it is called a vacancy. In a real crystal when atoms are out of line, it is called a dislocation.
6. Move the dish back and forth somewhat rapidly. (Don't spill the Bb's.) This simulates heating the metal. (Heating the metal gives the atoms more kinetic energy.)

A Particle Model of Metals: Atomic Bb's Objective: The objective of this lab is to learn more about the basic particle model for metals.

Review of Scientific Principles: Metallic crystals are not perfect. Sometimes there are empty spaces, vacancies, where an atom should be. There are also small mismatches, dislocations, in the rows of atoms, and these are found in all metals.

Applications: Defects in the crystal structure of metals control many of their properties including hardness and ductility.

Time: 40 minutes.

Materials and Supplies:

Plastic Petri dish.
Bb's.

Procedure:
Your team will be given a covered Petri dish containing Bb's. Keep the dish flat on the table, so no Bb's spill out! Answer the questions which correspond to a given procedure. Try to write complete idea answers that increases understanding.

1. The Bb's represent the atoms in a metal.
2. If you move the Petri dish back and forth, the atoms move. This simulates the movement of atoms in a metal when it is heated.
3. Move the dish back and forth and try to get the atoms arranged as neatly as possible. Slowing the motion of the dish and gradually stopping it simulates the formation of the crystal.
4. Make a sketch below showing how the Bb's are arranged. You don't have to draw all the Bb's!
5. Make a sketch below showing the arrangement of Bb's around an empty space. When this happens in the metal, it is called a vacancy. In a real crystal when atoms are out of line, it is called a dislocation.
6. Move the dish back and forth somewhat rapidly. (Don't spill the Bb's.) This simulates heating the metal. (Heating the metal gives the atoms more kinetic energy.)

From: MAST Handbook - http://matse1.matse.illinois.edu/metals/b.html

Questions:

1. Describe the bonding between the atoms in a metal.
2. What type of energy do "moving" atoms possess?
3. Do the atoms in a crystal move?
4. Describe the arrangement of the Bb's. Are there any empty spaces, i.e., places where a Bb is missing?
5. Are the Bb's arranged perfectly? Would you expect atoms to be perfectly arranged?
6. Do more or less defects exist in the metal when it is heated?

Chapter 11 Handout 1

Ceramics: Optical Properties

Name____________________ **Date**______________

1. What would be the critical angle of incidence for light transmission from a single-crystal Al_2O_3 (n = 1.76) into a coating of silica (n = 1.458)?

2. Lead Oxide has a refractive index of 2.61, while silica glass has a refractive index of 1.458. Lead oxide is sometimes added to Silica Glass for decorative purposes. Why?

Chapter 14 Handout 1

Chemistry and Classification of Polymers

Name____________________ Date_____________

1. Chain addition polymerization is when a molecule is added to a polymer one molecule at a time. Describe the three main steps of this polymerization mechanism.

2. The degree of polymerization can be defined as the molecular weight of the polymer divided by the molecular weight of the monomer unit. Take a sample of polyvinyl chloride. Shown is the structure of the monomer unit of this polymer –vinyl chloride (C2H3Cl).

a) Calculate the molecular weight of the monomer unit – vinyl chloride.

b) The number of average molecular weight for this polymer is 21,150 g/mol. Calculate the degree of polymerization. Describe what this value indicates.

Chapter 14 Handout 2

SCIENCE AND TECHNOLOGY OF MATERIALS

Student Questions: Reading Assignments on Teflon

Below are some questions for students based on the reading assignments on Teflon:

1. In Teflon, the hydrogens in polyethylene are replaced by which element?

2. Describe the properties of Teflon that make it useful in non-stick cooking pans.

3. In addition to nonstick pans, name two other applications for Teflon discussed in the chapter.

4. What product did Wilbert L. Gore eventually develop after leaving DuPont?

Appendix

Standards

The material in the following pages are the standards that were used to support the reading material. They are based on the NGSS (New Generation Science Standards) and Common Cord.

Chapter 1

[92]Standards

Standards met in the chapter based on **NGSS (New Generation Science Standards)** are:

MS-PS1-5 – Develop and use a model to describe how the total number of atoms does not change a chemical reaction and thus mass is conserved [Classification Statement: Emphasis is on the conservation of Matter and on physical models and drawings, including digital forms that represent atoms] [Assessment Boundary: Assessment does not include the use of atomic masses, balancing atomic equations, or intermolecular forces].

MS-ESS3-1 – Construct a scientific explanation based on evidence for how the uneven distribution of earth's minerals, energy, and groundwater resources are the result of past and current geological processes. [Clarification Statement: Emphasis is on how these resources are limited and typically non-renewable, and how their distributions are significantly changing as a result of removal by humans. Examples of uneven distribution of resources as a result of past processes include but are not limited to petroleum (locations of the burial of organic marine sediment and subsequent geological traps), metal ores (location of past volcanic and hydrothermal activity associated with subduction zones), and soil (location of active weathering and/or disposition of rock).]

R. A. Wesolowski et al., *The World of Materials*,
https://doi.org/10.1007/978-3-030-17847-5

HS-PS1-4 – Develop a model to illustrate that the release of absorption of energy from a chemical reaction system depends upon the change in total bond energy. [Clarification Statement: Emphasis is on the idea that a chemical reaction is a system that affects the energy change. Examples of modes could include molecular-level drawings and diagrams of reactions and graphs showing the relative energies of reactants and products, and representations showing energy is conserved.] [Assessment Boundary: Assessment does not include calculating the total bond energy changes during a chemical reaction from the bond energies of reactants and products.]

HS-PS2-4 – Use mathematical representation of Newton's Law of Gravitational Attraction and Coulomb's Law to describe and predict gravitational and electrostatic forces between objects. [Clarification Statement: Emphasis is on both the quantitative and the conceptual descriptions of gravitational and electric field.] [Assessment is limited to systems with two objects.]

HS-PS2-6 – Communicate scientific and technical information about why the molecular-level structure is important in the functioning of designed materials. [Clarification Statement: Emphasis is on the attractive and repulsive forces that determine the functioning of the material. Examples include why electrically conductive materials are often made of metal, flexible but durable materials are made up of long chained molecules, and pharmaceuticals are designed to interact with specific receptors.] [Assessment Boundary: Assessment is limited to provided molecular structures of specific designed materials.]

Science and engineering practices	Disciplinary core ideas	Crosscutting concepts
Developing and using models Modeling 9–12 builds on K-8 and progresses to using, synthesizing, and developing models to predict and show relationships among variables between systems and their components in the natural and designed worlds **Using mathematics and computational thinking** Using mathematical representation of phenomena to describe explanations (HS-PS2-2, HS-PS2-4)	**PS1.A structure and properties of matter** Each atom has a charged substructure consisting of a nucleus, which is made of protons and neutrons, surrounded by electrons (HS-PS1-1) A stable molecule has less energy that the same set of atoms separated; one must provide at least this energy in order to take the molecule apart (HS-PS1-4) The structure and interaction of matter at the bulk scale are determined by electrical forces within and between atoms. (secondary to HS-PS2-6)	**Energy and matter** Changes of energy and matter in a system can be described in terms of energy and matter flows into, out of, and within that system (HS-PS1-6) **Patterns** Different patterns may be observed to each of the scales at which a system is studied and can provide evidence for causality in explanations of phenomena (HS-PS2-4) **Structure and function** Investigating or designing new systems or structures requires a detailed examination of the properties of different materials, the structures of different components, and connections of components to reveal its function and/or solve a problem (HS-PS2-6)

[93]Common Core Standards

CCSS.ELA-LITERACY.RST.11-12.9.8
Evaluate the hypotheses, data, analysis, and conclusions in a science or technical text, verifying the data when possible and corroborating or challenging conclusions with other sources of information.

CCSS.ELA-LITERACY.RST.11-12.9
Synthesize information from a range of sources (e.g., texts, experiments, simulations) into a coherent understanding of a process, phenomenon, or concept, resolving conflicting information when possible.
Range of Reading and Level of Text Complexity:

CCSS.ELA-LITERACY.RST.11-12.10
By the end of grade 12, read and comprehend science/technical texts in the grades 11-CCR text complexity band independently and proficiently.

CCSS.ELA-LITERACY.RST.11-12.9
Synthesize information from a range of sources (e.g., texts, experiments, simulations) into a coherent understanding of a process, phenomenon, or concept, resolving conflicting information when possible.

CCSS.ELA-LITERACY.RST.6–8.10
By the end of grade 8, read and comprehend science/technical texts in the grades 6–8 text complexity band independently and proficiently.

Chapter 2

[92]Standards

Standards met in the chapter based on **NGSS (New Generation Science Standards)** are:

MS-PS1-1 – Develop models to describe the atomic composition of simple molecules and extended structures. [Clarification Statement: Emphasis is on developing models of molecules that vary in complexity. Examples of simple molecules could include ammonia and methanol. Examples of extended structures could include sodium chloride or diamonds. Examples of molecular-level models could include drawings, 3D ball, and stick structures, or computer representations showing different molecules with different types of atoms.] [Assessment Boundary: Assessment does not include valence electrons and bonding energy, discussing the ionic nature of subunits of complex structures, or a complete depiction of all individual atoms in a complex molecule or extended structure.]

MS-PS1-2 – Analyze and interpret data on the properties of substances before and after the substances interact to determine if a chemical reaction has occurred. [Clarification Statement: Examples of reactions could include burning sugar or steel wool, fat reacting with sodium hydroxide, and mixing zinc with hydrogen chloride.] [Assessment Boundary: Assessment is limited to analysis of the following properties: density, melting point, boiling point, solubility, flammability, and odor.]

HS-PS1-1 – Use the periodic table as a model to predict the relative properties of elements based on the patterns of electrons in the outermost energy level

of atoms. [Clarification Statement: Examples of properties that could be predicted from patterns could include reactivity of metals, types of bonds formed, numbers of bonds formed, and reactions with oxygen.] [Assessment Boundary: Assessment is limited to main group elements. Assessment does not include quantitative understanding of ionization energy beyond relative trends.]

HS-PS1-2 – Construct and revise an explanation for the outcome of a simple chemical reaction based on the outermost electron states of atoms, trends in the periodic table, and knowledge of the patterns of chemical properties. [Clarification Statement: Examples of chemical reactions could include the reaction of sodium and chlorine, of carbon and oxygen, or of carbon and hydrogen.] [Assessment Boundary: Assessment is limited to chemical reactions involving main group elements and combustion reactions.]

HS-PS1-4 – Develop a model to illustrate that the release of absorption of energy from a chemical reaction system depends upon the change in total bond energy. [Clarification Statement: Emphasis is on the idea that a chemical reaction is a system that affects the energy change. Examples of modes could include molecular-level drawings and diagrams of reactions and graphs showing the relative energies of reactants and products, and representations showing energy is conserved.] [Assessment Boundary: Assessment does not include calculating the total bond energy changes during a chemical reaction from the bond energies of reactants and products.]

HS-PS1-8 – Develop models to illustrate the changes in the composition of the nucleus of the atom and the energy released during the processes of fission, fusion, and radioactive decay. [Clarification Statement: Emphasis is on simple qualitative models, such as pictures or diagrams, and on the scale of energy released in nuclear processes relative to other kinds of transformations.] [Assessment Boundary: Assessment does not include quantitative calculation of energy released. Assessment is limited to alpha, beta, and gamma radioactive decays.]

c[93]Common Core Standards

CCSS.ELA-LITERACY.RST.11-12.9.8
Evaluate the hypotheses, data, analysis, and conclusions in a science or technical text, verifying the data when possible and corroborating or challenging conclusions with other sources of information.
Range of Reading and Level of Text Complexity:

CCSS.ELA-LITERACY.RST.11-12.10
By the end of grade 12, read and comprehend science/technical texts in the grades 11-CCR text complexity band independently and proficiently.

CCSS.ELA-LITERACY.RST.6-8.10
By the end of grade 8, read and comprehend science/technical texts in the grades 6–8 text complexity band independently and proficiently.

CCSS.ELA-LITERACY.RST.6-8.4
Determine the meaning of symbols, key terms, and other domain-specific words and phrases as they are used in a specific scientific or technical context relevant to grades 6–8 texts and topics.

Science and engineering practices	Disciplinary core ideas	Crosscutting concepts
Developing and using models Modeling in 6–8 builds on K–5 and progresses to developing, using, and revising models to describe, test, and predict more abstract phenomena and design systems. Develop a model to predict and/or describe phenomena (MS-PS1-1, MS-PS1-4) **Analyzing and interpreting data** Analyzing data in 6–8 builds on K–5 and progresses to extending quantitative analysis to investigations, distinguishing between correlation and causation, and basic statistical techniques of data and error analysis Analyze and interpret data to determine similarities and differences in findings (MS-PS1-2) Modeling in 9–12 builds on K–8 and progresses to using, synthesizing, and developing models to predict and show relationships among variables between systems and their components in the natural and designed worlds Develop a model based on evidence to illustrate the relationships between systems or between components of a system (HS-PS1-4, HS-PS1-8) Use a model to predict the relationships between systems or between components of a system (HS-PS1-1)	**PS1.A: Structure and properties of matter** Substances are made from different types of atoms, which combine with one another in various ways. Atoms form molecules that range in size from two to thousands of atoms (MS-PS1-1) Each pure substance has characteristic physical and chemical properties (for any bulk quantity under given conditions) that can be used to identify it (MS-PS1-2, MS-PS1-3) PS1.A: Structure and properties of matter Each atom has a charged substructure consisting of a nucleus, which is made of protons and neutrons, surrounded by electrons (HS-PS1-1) The periodic table orders elements horizontally by the number of protons in the atom's nucleus and places those with similar chemical properties in columns. The repeating patterns of this table reflect patterns of outer electron states (HS-PS1-1, HS-PS1-2) The structure and interactions of matter at the bulk scale are determined by electrical forces within and between atoms (HS-PS1-3) A stable molecule has less energy than the same set of atoms separated; one must provide at least this energy in order to take the molecule apart (HS-PS1-4)	**Patterns** Macroscopic patterns are related to the nature of microscopic and atomic-level structure (MS-PS1-2) Different patterns may be observed at each of the scales at which a system is studied and can provide evidence for causality in explanations of phenomena (HS-PS1-1, HS-PS1-2, HS-PS1-3, HS- PS1-5) **Energy and matter** In nuclear processes, atoms are not conserved, but the total number of protons plus neutrons is conserved (HS-PS1-8) The total amount of energy and matter in closed systems is conserved (HS-PS1-7) Changes of energy and matter in a system can be described in terms of energy and matter flows into, out of, and within that system (HS-PS1-4) **Stability and change** Much of science deals with constructing explanations of how things change and how they remain stable (HS-PS1-6)

Chapter 3

[92]Standards

Standards met in the chapter based on **NGSS (New Generation Science Standards)** are:

MS-PS1-2 – Analyze and interpret data on the properties of substances before and after the substances interact to determine if a chemical reaction has occurred. [Clarification Statement: Examples of reactions could include burning sugar or steel wool, fat reacting with sodium hydroxide, and mixing zinc with hydrogen chloride.] [Assessment Boundary: Assessment is limited to analysis of the following properties: density, melting point, boiling point, solubility, flammability, and odor.]

MS-PS1-3 – Gather and make sense of information to describe that synthetic materials come from natural resources and impact society. [Clarification Statement: Emphasis is on natural resources that undergo a chemical process to form the synthetic material. Examples of new materials could include new medicine, foods, and alternative fuels.] [Assessment Boundary: Assessment is limited to qualitative information.]

MS-PS1-5 – Develop and use a model to describe how the total number of atoms does not change in a chemical reaction and thus mass is conserved. [Clarification Statement: Emphasis is on law of conservation of matter and on physical models or drawings, including digital forms, that represent atoms.] [Assessment Boundary: Assessment does not include the use of atomic masses, balancing symbolic equations, or intermolecular forces.]

HS-PS1-1 – Use the periodic table as a model to predict the relative properties of elements based on the patterns of electrons in the outermost energy level of atoms. [Clarification Statement: Examples of properties that could be predicted from patterns could include reactivity of metals, types of bonds formed, numbers of bonds formed, and reactions with oxygen.] [Assessment Boundary: Assessment is limited to main group elements. Assessment does not include quantitative understanding of ionization energy beyond relative trends.]

HS-PS1-4 – Develop a model to illustrate that the release of absorption of energy from a chemical reaction system depends upon the change in total bond energy. [Clarification Statement: Emphasis is on the idea that a chemical reaction is a system that affects the energy change. Examples of modes could include molecular-level drawings and diagrams of reactions and graphs showing the relative energies of reactants and products, and representations showing energy is conserved.] [Assessment Boundary: Assessment does not include calculating the total bond energy changes during a chemical reaction from the bond energies of reactants and products.]

HS-PS1-7 – Use mathematical representations to support the claim that atoms, and therefore mass, are conserved during a chemical reaction. [Clarification Statement: Emphasis is on using mathematical ideas to communicate the proportional relationships between masses of atoms in the reactants and the products, and the translation of these relationships to the macroscopic scale using the mole as the conversion from the atomic to the macroscopic scale. Emphasis is on assessing students' use of mathematical thinking and not on memorization and rote application of problem-solving techniques.] [Assessment Boundary: Assessment does not include complex chemical reactions.]

Science and engineering practices	Disciplinary core ideas	Crosscutting concepts
Analyzing and interpreting data Analyzing data in 6–8 builds on K–5 and progresses to extending quantitative analysis to investigations, distinguishing between correlation and causation, and basic statistical techniques of data and error analysis Analyze and interpret data to determine similarities and differences in findings (MS-PS1-2) **Obtaining, evaluating, and communicating information** Obtaining, evaluating, and communicating information in 6–8 builds on K–5 and progresses to evaluating the merit and validity of ideas and methods Gather, read, and synthesize information from multiple appropriate sources and assess the credibility, accuracy, and possible bias of each publication and methods used, and describe how they are supported or not supported by evidence (MS-PS1-3) Modeling in 9–12 builds on K–8 and progresses to using, synthesizing, and developing models to predict and show relationships among variables between systems and their components in the natural and designed worlds Use a model to predict the relationships between systems or between components of a system (HS-PS1-1)	**PS1.A: Structure and properties of matter** Each pure substance has characteristic physical and chemical properties (for any bulk quantity under given conditions) that can be used to identify it (MS-PS1-2, MS-PS1-3) Each atom has a charged substructure consisting of a nucleus, which is made of protons and neutrons, surrounded by electrons (HS-PS1-1) A stable molecule has less energy than the same set of atoms separated; one must provide at least this energy in order to take the molecule apart (HS-PS1-4) **PS1.B: Chemical reactions** Substances react chemically in characteristic ways. In a chemical process, the atoms that make up the original substances are regrouped into different molecules, and these new substances have different properties from those of the reactants (MS-PS1-2, MS-PS1-3, MS-PS1-5) The total number of each type of atom is conserved, and thus the mass does not change (MS-PS1-5)	**Patterns** Macroscopic patterns are related to the nature of microscopic and atomic-level structure (MS-PS1-2) Different patterns may be observed at each of the scales at which a system is studied and can provide evidence for causality in explanations of phenomena (HS-PS1-1, HS-PS1-2, HS-PS1-3, HS-PS1-5) **Energy and matter** The transfer of energy can be tracked as energy flows through a designed or natural system (MS-PS1-6) In nuclear processes, atoms are not conserved, but the total number of protons plus neutrons is conserved (HS-PS1-8) **Structure and function** Structures can be designed to serve particular functions by taking into account properties of different materials, and how materials can be shaped and used (MS-PS1-3)

HS.PS3.A (HS-PS2-4),(HS-PS2-5); **HS.PS3.C** (HS-PS2-1); **HS.PS4.B** (HS-PS2-5); **HS. ESS1.A** (HS-PS2-1),(HS-PS2-2),(HS-PS2-4); **HS.ESS1.B** (HS-PS2-4); **HS.ESS1.C** (HS-PS2-1),(HS-PS2-2),(HS-PS2-4); **HS.ESS2.A** (HS-PS2-5); **HS.ESS2.C** (HS-PS2-1),(HS-PS2-4); **HS.ESS3.A** (HS-PS2-4),(HS-PS2-5)
MS.PS1.A (HS-PS2-6); **MS.PS2.A** (HS-PS2-1),(HS-PS2-2),(HS-PS2-3); **MS.PS2.B** (HS-PS2-4),(HS-PS2-5),(HS-PS2-6); **MS.PS3.C** (HS-PS2-1),(HS-PS2-2),(HS-PS2-3); **MS.ESS1.B** (HS-PS2-4),(HS-PS2-5)

RST.11-12.1	Cite specific textual evidence to support analysis of science and technical texts, attending to important distinctions the author makes and to any gaps or inconsistencies in the account
RST.11-12.7	Integrate and evaluate multiple sources of information presented in diverse formats and media (e.g., quantitative data, video, multimedia) in order to address a question or solve a problem (HS-PS2-1)
WHST.9-12.2	Write informative/explanatory texts, including the narration of historical events, scientific procedures experiments, or technical processes
WHST.9-12.7	Conduct short as well as more sustained research projects to answer a question (including a self-generated question) or solve a problem; narrow or broaden the inquiry when appropriate; synthesize multiple sources on the subject, demonstrating understanding of the subject under investigation (HS-PS2-3),(HS-PS2-5)
WHST.11-12.8	Gather relevant information from multiple authoritative print and digital sources, using advanced searches effectively; assess the strengths and limitations of each source in terms of the specific task, purpose, and audience; integrate information into the text selectively to maintain the flow of ideas, avoiding plagiarism and overreliance on any one source and following a standard format for citation
WHST.9-12.9	Draw evidence from informational texts to support analysis, reflection, and research
MP.2	Reason abstractly and quantitatively (HS-PS2-1, HS-PS2-2, HS-PS2-4)
MP.4	Model with mathematics (HS-PS2-1, HS-PS2-2, HS-PS2-4)
HSN-Q.A.1	Use units as a way to understand problems and to guide the solution of multi-step problems; choose and interpret units consistently in formulas; choose and interpret the scale and the origin in graphs and data displays (HS-PS2-1, HS-PS2-2, HS-PS2-4)
HSN-Q.A.2	Define appropriate quantities for the purpose of descriptive modeling
HSN-Q.A.3	Choose a level of accuracy appropriate to limitations on measurement when reporting quantities (HS-PS2-1, HS-PS2-2, HS-PS2-4)
HSA-SSE.A.1	Interpret expressions that represent a quantity in terms of its context (HS-PS2-1, HS-PS2-4)
HSA-SSE.B.3	Choose and produce an equivalent form of an expression to reveal and explain properties of the quantity represented by the expression (HS-PS2-1, HS-PS2-4)
HSA-CED.A.1	Create equations and inequalities in one variable and use them to solve problems (HS-PS2-1, HS-PS2-2)
HSA-CED.A.2	Create equations in two or more variables to represent relationships between quantities; graph equations on coordinate axes with labels and scales.
HSA-CED.A.4	Rearrange formulas to highlight a quantity of interest, using the same reasoning as in solving equations.
HSF-IF.C.7	Graph functions expressed symbolically and show key features of the graph, by in hand in simple cases and using technology for more complicated cases
HSS-ID.A.1	Represent data with plots on the real number line (dot plots, histograms, and box plots)

[93]Common Core Standards

Range of Reading and Level of Text Complexity:

CCSS.ELA-LITERACY.RST.11-12.10
By the end of grade 12, read and comprehend science/technical texts in the grades 11-CCR text complexity band independently and proficiently.

CCSS.ELA-LITERACY.RST.6-8.7
Integrate quantitative or technical information expressed in words in a text with a version of that information expressed visually (e.g., in a flowchart, diagram, model, graph, or table).

CCSS.ELA-LITERACY.RST.6-8.8
Distinguish among facts, reasoned judgment based on research findings, and speculation in a text.

Chapter 4

[92]Standards

Standards met in the chapter based on **NGSS (New Generation Science Standards)** are:

MS-PS1-1. Develop models to describe the atomic composition of simple molecules and extended structures. [Clarification Statement: Emphasis is on developing models of molecules that vary in complexity. Examples of simple molecules could include ammonia and methanol. Examples of extended structures could include sodium chloride or diamonds. Examples of molecular-level models could include drawings, 3D ball, and stick structures, or computer representations showing different molecules with different types of atoms.] [Assessment Boundary: Assessment does not include valence electrons and bonding energy, discussing the ionic nature of subunits of complex structures, or a complete depiction of all individual atoms in a complex molecule or extended structure.]

MS-PS2-5. Conduct an investigation and evaluate the experimental design to provide evidence that fields exist between objects exerting forces on each other even though the objects are not in contact. [Clarification Statement: Examples of this phenomenon could include the interactions of magnets, electrically-charged strips of tape, and electrically-charged pith balls. Examples of investigations could include first-hand experiences or simulations.] [Assessment Boundary: Assessment is limited to electric and magnetic fields, and limited to qualitative evidence for the existence of fields.]

HS-PS1-7. Use mathematical representations to support the claim that atoms, and therefore mass, are conserved during a chemical reaction. [Clarification Statement: Emphasis is on using mathematical ideas to communicate the proportional relationships between masses of atoms in the reactants and the products, and the translation of these relationships to the macroscopic scale using the mole as the conversion from the atomic to the macroscopic scale. Emphasis is on assessing students' use of mathematical thinking and not on memorization and rote application of problem-solving techniques.] [Assessment Boundary: Assessment does not include complex chemical reactions.]

Science and engineering practices	Disciplinary core ideas	Crosscutting concepts
Planning and carrying out investigations Planning and carrying out investigations to answer questions or test solutions to problems in 6–8 builds on K–5 experiences and progresses to include investigations that use multiple variables and provide evidence to support explanations or design solutions Plan an investigation individually and collaboratively, and in the conduct an investigation and evaluate the experimental design to produce data to serve as the basis for evidence that can meet the goals of the investigation (MS-PS2-5). **Obtaining, evaluating, and communicating information** Obtaining, evaluating, and communicating information in 6–8 builds on K–5 and progresses to evaluating the merit and validity of ideas and methods Gather, read, and synthesize information from multiple appropriate sources and assess the credibility, accuracy, and possible bias of each publication and methods used, and describe how they are supported or not supported by evidence (MS-PS1-3) Modeling in 9–12 builds on K–8 and progresses to using, synthesizing, and developing models to predict and show relationships among variables between systems and their components in the natural and designed worlds Use a model to predict the relationships between systems or between components of a system (HS-PS1-1)	**PS1.A: Structure and properties of matter** Each pure substance has characteristic physical and chemical properties (for any bulk quantity under given conditions) that can be used to identify it (MS-PS1-2, MS-PS1-3) Each atom has a charged substructure consisting of a nucleus, which is made of protons and neutrons, surrounded by electrons (HS-PS1-1) A stable molecule has less energy than the same set of atoms separated; one must provide at least this energy in order to take the molecule apart (HS-PS1-4) The total number of each type of atom is conserved, and thus the mass does not change (MS-PS1-5) **PS2.B: Types of interactions** Forces that act at a distance (electric, magnetic, and gravitational) can be explained by fields that extend through space and can be mapped by their effect on a test object (a charged object, or a ball, respectively) (MS-PS2-5)	**Cause and effect** Cause and effect relationships may be used to predict phenomena in natural or designed systems (MS-PS2-3, MS-PS2-5). **Patterns** Macroscopic patterns are related to the nature of microscopic and atomic-level structure (MS-PS1-2) Different patterns may be observed at each of the scales at which a system is studied and can provide evidence for causality in explanations of phenomena (HS-PS1-1, HS-PS1-2, HS-PS1-3, HS- PS1-5) **Structure and function** Structures can be designed to serve particular functions by taking into account properties of different materials, and how materials can be shaped and used (MS-PS1-3)

HS.PS3.A (HS-PS2-4),(HS-PS2-5); **HS.PS3.C** (HS-PS2-1); **HS.PS4.B** (HS-PS2-5); **HS. ESS1.A** (HS-PS2-1),(HS-PS2-2),(HS-PS2-4); **HS.ESS1.B** (HS-PS2-4); **HS.ESS1.C** (HS-PS2-1),(HS-PS2-2),(HS-PS2-4); **HS.ESS2.A** (HS-PS2-5); **HS.ESS2.C** (HS-PS2-1),(HS-PS2-4); **HS.ESS3.A** (HS-PS2-4),(HS-PS2-5)
MS.PS1.A (HS-PS2-6); **MS.PS2.A** (HS-PS2-1),(HS-PS2-2),(HS-PS2-3); **MS.PS2.B** (HS-PS2-4),(HS-PS2-5),(HS-PS2-6); **MS.PS3.C** (HS-PS2-1),(HS-PS2-2),(HS-PS2-3); **MS. ESS1.B** (HS-PS2-4),(HS-PS2-5)

RST.11-12.1	Cite specific textual evidence to support analysis of science and technical texts, attending to important distinctions the author makes and to any gaps or inconsistencies in the account
RST.11-12.7	Integrate and evaluate multiple sources of information presented in diverse formats and media (e.g., quantitative data, video, multimedia) in order to address a question or solve a problem (HS-PS2-1)
WHST.9-12.2	Write informative/explanatory texts, including the narration of historical events, scientific procedures/ experiments, or technical processes
WHST.9-12.7	Conduct short as well as more sustained research projects to answer a question (including a self-generated question) or solve a problem; narrow or broaden the inquiry when appropriate; synthesize multiple sources on the subject, demonstrating understanding of the subject under investigation (HS-PS2-3, HS-PS2-5)
WHST.11-12.8	Gather relevant information from multiple authoritative print and digital sources, using advanced searches effectively; assess the strengths and limitations of each source in terms of the specific task, purpose, and audience; integrate information into the text selectively to maintain the flow of ideas, avoiding plagiarism and overreliance on any one source and following a standard format for citation
WHST.9-12.9	Draw evidence from informational texts to support analysis, reflection, and research
MP.2	Reason abstractly and quantitatively (HS-PS2-1, HS-PS2-2, HS-PS2-4)
MP.4	Model with mathematics (HS-PS2-1, HS-PS2-2, HS-PS2-4)
HSN-Q.A.1	Use units as a way to understand problems and to guide the solution of multi-step problems; choose and interpret units consistently in formulas; choose and interpret the scale and the origin in graphs and data displays (HS-PS2-1, HS-PS2-2, HS-PS2-4)
HSN-Q.A.2	Define appropriate quantities for the purpose of descriptive modeling
HSN-Q.A.3	Choose a level of accuracy appropriate to limitations on measurement when reporting quantities (HS-PS2-1, HS-PS2-2, HS-PS2-4)
HSA-SSE.A.1	Interpret expressions that represent a quantity in terms of its context (HS-PS2-1, HS-PS2-4)
HSA-SSE.B.3	Choose and produce an equivalent form of an expression to reveal and explain properties of the quantity represented by the expression (HS-PS2-1, HS-PS2-4)
HSA-CED.A.1	Create equations and inequalities in one variable and use them to solve problems (HS-PS2-1, HS-PS2-2)
HSA-CED.A.2	Create equations in two or more variables to represent relationships between quantities; graph equations on coordinate axes with labels and scales
HSA-CED.A.4	Rearrange formulas to highlight a quantity of interest, using the same reasoning as in solving equations
HSF-IF.C.7	Graph functions expressed symbolically and show key features of the graph, by in hand in simple cases and using technology for more complicated cases
HSS-ID.A.1	Represent data with plots on the real number line (dot plots, histograms, and box plots)

[93]Common Core Standards

Range of Reading and Level of Text Complexity:

CCSS.ELA-LITERACY.RST.11-12.10
By the end of grade 12, read and comprehend science/technical texts in the grades 11-CCR text complexity band independently and proficiently.

CCSS.ELA-LITERACY.RST.6-8.7
Integrate quantitative or technical information expressed in words in a text with a version of that information expressed visually (e.g., in a flowchart, diagram, model, graph, or table).

CCSS.ELA-LITERACY.RST.6-8.8
Distinguish among facts, reasoned judgment based on research findings, and speculation in a text.

Chapter 5

[92]Standards

Standards met in the chapter based on **NGSS (New Generation Science Standards)** are:

MS-PS2-1 – Apply Newton's Third Law to design a solution to a problem involving the motion of two colliding objects.∗ [Clarification Statement: Examples of practical problems could include the impact of collisions between two cars, between a car and stationary objects, and between a meteor and a space vehicle.] [Assessment Boundary: Assessment is limited to vertical or horizontal interactions in one dimension.]

MS-PS2-2 – Plan an investigation to provide evidence that the change in an object's motion depends on the sum of the forces on the object and the mass of the object. [Clarification Statement: Emphasis is on balanced (Newton's First Law) and unbalanced forces in a system, qualitative comparisons of forces, mass and changes in motion (Newton's Second Law), frame of reference, and specification of units.] [Assessment Boundary: Assessment is limited to forces and changes in motion in one dimension in an inertial reference frame and to change in one variable at a time. Assessment does not include the use of trigonometry.]

MS-PS2-3 – Ask questions about data to determine the factors that affect the strength of electric and magnetic forces. [Clarification Statement: Examples of devices that use electric and magnetic forces could include electromagnets, electric motors, or generators. Examples of data could include the effect of the number of turns of wire on the strength of an electromagnet, or the effect of increasing the number or strength of magnets on the speed of an electric motor.] [Assessment Boundary: Assessment about questions that require quantitative answers is limited to proportional reasoning and algebraic thinking.]

HS-PS1-1 – Use the periodic table as a model to predict the relative properties of elements based on the patterns of electrons in the outermost energy level of atoms. [Clarification Statement: Examples of properties that could be predicted from patterns could include reactivity of metals, types of bonds formed, numbers of bonds formed, and reactions with oxygen.] [Assessment Boundary: Assessment is limited to main group elements. Assessment does not include quantitative understanding of ionization energy beyond relative trends.]

HS-PS1-3 – Plan and conduct an investigation to gather evidence to compare the structure of substances at the bulk scale to infer the strength of electrical forces between particles. [Clarification Statement: Emphasis is on understanding the strengths of forces between particles, not on naming specific intermolecular forces (such as dipole-dipole). Examples of particles could include ions, atoms, molecules, and networked materials (such as graphite). Examples of bulk properties of substances could include the melting point and boiling point, vapor pressure, and surface tension.] [Assessment Boundary: Assessment does not include Raoult's law calculations of vapor pressure.]

HS-PS1-4 – Develop a model to illustrate that the release or absorption of energy from a chemical reaction system depends upon the changes in total bond energy. [Clarification Statement: Emphasis is on the idea that a chemical reaction is a system that affects the energy change. Examples of models could include molecular-level drawings and diagrams of reactions, graphs showing the relative energies of reactants and products, and representations showing energy is conserved.] [Assessment Boundary: Assessment does not include calculating the total bond energy changes during a chemical reaction from the bond energies of reactants and products.]

HS-PS1-5 – Apply scientific principles and evidence to provide an explanation about the effects of changing the temperature or concentration of the reacting particles on the rate at which a reaction occurs. [Clarification Statement: Emphasis is on student reasoning that focuses on the number and energy of collisions between molecules.] [Assessment Boundary: Assessment is limited to simple reactions in which there are only two reactants; evidence from temperature, concentration, and rate data; and qualitative relationships between rate and temperature.]

Science and engineering practices	Disciplinary core ideas	Crosscutting concepts
Asking questions and defining problems Asking questions and defining problems in grades 6–8 builds from grades K–5 experiences and progresses to specifying relationships between variables, and clarifying arguments and models Ask questions that can be investigated within the scope of the classroom, outdoor environment, and museums and other public facilities with available resources and, when appropriate, frame a hypothesis based on observations and scientific principles (MS-PS2-3) Planning and carrying out investigations Planning and carrying out investigations to answer questions or test solutions to problems in 6–8 builds on K–5 experiences and progresses to include investigations that use multiple variables and provide evidence to support explanations or design solutions. Plan an investigation individually and collaboratively, and in the design: Identify independent and dependent variables and controls, what tools are needed to do the gathering, how measurements will be recorded, and how many data are needed to support a claim (MS-PS2-2) **Planning and carrying out investigations** Planning and carrying out investigations in 9–12 builds on K-8 experiences and progresses to include investigations that provide evidence for and test conceptual, mathematical, physical, and empirical models Plan and conduct an investigation individually and collaboratively to produce data to serve as the basis for evidence, and in the design: Decide on types, how much, and accuracy of data needed to produce reliable measurements and consider limitations on the precision of the data (e.g., number of trials, cost, risk, time), and refine the design accordingly (HS-PS1-3) **Constructing explanations and designing solutions** Constructing explanations and designing solutions in 6–8 builds on K–5 experiences and progresses to include constructing explanations and designing solutions supported by multiple sources of evidence consistent with scientific ideas, principles, and theories Apply scientific ideas or principles to design an object, tool, process or system (MS-PS2-1) **Developing and using models** Modeling in 9–12 builds on K–8 and progresses to using, synthesizing, and developing models to predict and show relationships among variables between systems and their components in the natural and designed worlds Use a model to predict the relationships between systems or between components of a system (HS-PS1-1)	**PS2.A: Forces and motion** For any pair of interacting objects, the force exerted by the first object on the second object is equal in strength to the force that the second object exerts on the first, but in the opposite direction (Newton's third law) (MS-PS2-1) The motion of an object is determined by the sum of the forces acting on it; if the total force on the object is not zero, its motion will change. The greater the mass of the object, the greater the force needed to achieve the same change in motion. For any given object, a larger force causes a larger change in motion (MS-PS2-2). All positions of objects and the directions of forces and motions must be described in an arbitrarily chosen reference frame and arbitrarily chosen units of size. In order to share information with other people, these choices must also be shared (MSPS2-2). **PS1.A: Structure and properties of matter** Each atom has a charged substructure consisting of a nucleus, which is made of protons and neutrons, surrounded by electrons (HS-PS1-1) The periodic table orders elements horizontally by the number of protons in the atom's nucleus and places those with similar chemical properties in columns. The repeating patterns of this table reflect patterns of outer electron states (HS-PS1–1, HS-PS1–2) The structure and interactions of matter at the bulk scale are determined by electrical forces within and between atoms (HS-PS1–3) A stable molecule has less energy than the same set of atoms separated; one must provide at least this energy in order to take the molecule apart. (HS-PS1-4)	**Cause and effect** Cause and effect relationships may be used to predict phenomena in natural or designed systems (MS-PS2-3, MS-PS2-5) **Systems and system models** Models can be used to represent systems and their interactions – Such as inputs, processes and outputs – and energy and matter flows within systems (MS-PS2-1, MS-PS2-4) **Stability and change** Explanations of stability and change in natural or designed systems can be constructed by examining the changes over time and forces at different scales (MS-PS2-2) **Patterns** Different patterns may be observed at each of the scales at which a system is studied and can provide evidence for causality in explanations of phenomena. (HS-PS1-1, HS-PS1-2, HS-PS1-3, HS-PS1-5) ***Connections to engineering, technology, and applications of science*** **Influence of science, engineering, and technology on society and the natural world** The uses of technologies and any limitations on their use are driven by individual or societal needs, desires, and values; by the findings of scientific research; and by differences in such factors as climate, natural resources, and economic conditions (MS-PS2-1)

<table>
<tr><th>Science and engineering practices</th><th>Disciplinary core ideas</th><th>Crosscutting concepts</th></tr>
<tr><td colspan="3">MS-PS2 Motion and Stability: Forces and Interactions *The performance expectations marked with an asterisk integrate traditional science content with engineering through a Practice or Disciplinary Core Idea. The section entitled "Disciplinary Core Ideas" is reproduced verbatim from A Framework for K-12 Science Education: Practices, Cross-Cutting Concepts, and Core Ideas. Integrated and reprinted with permission from the National Academy of Sciences.
RST.6-8.1 Cite specific textual evidence to support analysis of science and technical texts, attending to the precise details of explanations or descriptions (MS-PS2-1, MSPS2-3)
RST.6-8.3 Follow precisely a multistep procedure when carrying out experiments, taking measurements, or performing technical tasks. (MS-PS2-1, MS-PS2-2, MS-PS2-5)
WHST.6-8.1 Write arguments focused on discipline-specific content (MS-PS2-4)
WHST.6-8.7 Conduct short research projects to answer a question (including a self-generated question), drawing on several sources and generating additional related, focused questions that allow for multiple avenues of exploration (MS-PS2-1, MS-PS2-2, MS-PS2-5)
Mathematics:
MP.2 Reason abstractly and quantitatively (MS-PS2-1, MS-PS2-2, MS-PS2-3)
6.NS.C.5 Understand that positive and negative numbers are used together to describe quantities having opposite directions or values; use positive and negative numbers to represent quantities in real-world contexts, explaining the meaning of 0 in each situation (MS-PS2-1)
6.EE.A.2 Write, read, and evaluate expressions in which letters stand for numbers (MS-PS2-1, MS-PS2-2)
7.EE.B.3 Solve multi-step real-life and mathematical problems posed with positive and negative rational numbers in any form, using tools strategically. Apply properties of operations to calculate with numbers in any form; convert between forms as appropriate; and assess the reasonableness of answers using mental computation and estimation strategies (MS-PS2-1, MS-PS2-2)
7.EE.B.4 Use variables to represent quantities in a real-world or mathematical problem, and construct simple equations and inequalities to solve problems by reasoning about the quantities (MS-PS2-1, MS-PS2-2)</td></tr>
</table>

[93]Common Core Standards

CCSS.ELA-LITERACY.RST.11-12.4
Determine the meaning of symbols, key terms, and other domain-specific words and phrases as they are used in a specific scientific or technical context relevant to *grades 11–12 texts and topics.*

CCSS.ELA-LITERACY.RST.11-12.9
Synthesize information from a range of sources (e.g., texts, experiments, simulations) into a coherent understanding of a process, phenomenon, or concept, resolving conflicting information when possible.

CCSS.ELA LITERACY.RST.11 12.10
By the end of grade 12, read and comprehend science/technical texts in the grades 11-CCR text complexity band independently and proficiently.

CCSS.ELA-LITERACY.RST.6-8.4
Determine the meaning of symbols, key terms, and other domain-specific words and phrases as they are used in a specific scientific or technical context relevant to grades 6–8 texts and topics.

Chapter 6

[92]Standards

Standards met in the chapter based on **NGSS (New Generation Science Standards)** are:

MS-PS2-1. Apply Newton's Third Law to design a solution to a problem involving the motion of two colliding objects.* [Clarification Statement: Examples of practical problems could include the impact of collisions between two cars, between a car and stationary objects, and between a meteor and a space vehicle.] [Assessment Boundary: Assessment is limited to vertical or horizontal interactions in one dimension.]

MS-PS2-4. Construct and present arguments using evidence to support the claim that gravitational interactions are attractive and depend on the masses of interacting objects. [Clarification Statement: Examples of evidence for arguments could include data generated from simulations or digital tools; and charts displaying mass, strength of interaction, distance from the Sun, and orbital periods of objects within the solar system.] [Assessment Boundary: Assessment does not include Newton's Law of Gravitation or Kepler's Laws.]

MS-PS2-5. Conduct an investigation and evaluate the experimental design to provide evidence that fields exist between objects exerting forces on each other even though the objects are not in contact. [Clarification Statement: Examples of this phenomenon could include the interactions of magnets, electrically-charged strips of tape, and electrically-charged pith balls. Examples of investigations could include first-hand experiences or simulations.] [Assessment Boundary: Assessment is limited to electric and magnetic fields, and limited to qualitative evidence for the existence of fields.]

HS-PS1-1. Use the periodic table as a model to predict the relative properties of elements based on the patterns of electrons in the outermost energy level of atoms. [Clarification Statement: Examples of properties that could be predicted from patterns could include reactivity of metals, types of bonds formed, numbers of bonds formed, and reactions with oxygen.] [Assessment Boundary: Assessment is limited to main group elements. Assessment does not include quantitative understanding of ionization energy beyond relative trends.] [Assessment Boundary: techniques.] Assessment does not include complex chemical reactions.]

HS-PS1-3. Plan and conduct an investigation to gather evidence to compare the structure of substances at the bulk scale to infer the strength of electrical forces between particles. [Clarification Statement: Emphasis is on understanding the strengths of forces between particles, not on naming specific intermolecular forces (such as dipole-dipole). Examples of particles could include ions, atoms, molecules, and networked materials (such as graphite). Examples of bulk properties of substances could include the melting point and boiling point, vapor pressure, and surface tension.] [Assessment Boundary: Assessment does not include Raoult's law calculations of vapor pressure.]

Science and engineering practices	Disciplinary core ideas	Crosscutting concepts
Constructing explanations and designing solutions Constructing explanations and designing solutions in 6–8 builds on K–5 experiences and progresses to include constructing explanations and designing solutions supported by multiple sources of evidence consistent with scientific ideas, principles, and theories Apply scientific ideas or principles to design an object, tool, process, or system (MS-PS2-1) **Obtaining, evaluating, and communicating Information** Obtaining, evaluating, and communicating information in 6–8 builds on K–5 and progresses to evaluating the merit and validity of ideas and methods Gather, read, and synthesize information from multiple appropriate sources and assess the credibility, accuracy, and possible bias of each publication and methods used, and describe how they are supported or not supported by evidence (MS-PS1-3) Modeling in 9–12 builds on K–8 and progresses to using, synthesizing, and developing models to predict and show relationships among variables between systems and their components in the natural and designed worlds Use a model to predict the relationships between systems or between components of a system. (HS-PS1-1) **HS-PS1-1.** Use the periodic table as a model to predict the relative properties of elements based on the patterns of electrons in the outermost energy level of atoms. [clarification statement: Examples of properties that could be predicted from patterns could include reactivity of metals, types of bonds formed, numbers of bonds formed, and reactions with oxygen.] [assessment boundary: Assessment is limited to main group elements. Assessment does not include quantitative understanding of ionization energy beyond relative trends.] **HS-PS1-3. Planning and conducting investigations** Plan and conduct an investigation to gather evidence to compare the structure of substances at the bulk scale to infer the strength of electrical forces between particles. [Clarification statement: Emphasis is on understanding the strengths of forces between particles, not on naming specific intermolecular forces (such as dipole-dipole). Examples of particles could include ions, atoms, molecules, and networked materials (such as graphite). Examples of bulk properties of substances could include the melting point and boiling point, vapor pressure, and surface tension.] [assessment boundary: Assessment does not include Raoult's law calculations of vapor pressure.]	**PS1: Properties and structure of matter** Each pure substance has characteristic physical and chemical properties (for any bulk quantity under given conditions) that can be used to identify it (MS-PS1-2, MS-PS1-3) Each atom has a charged substructure consisting of a nucleus, which is made of protons and neutrons, surrounded by electrons (HS-PS1-1) **PS1.B: Chemical reactions** Substances react chemically in characteristic ways. In a chemical process, the atoms that make up the original substances are regrouped into different molecules, and these new substances have different properties from those of the reactants (MS-PS1-2, MS-PS1-3, MS-PS1-5) The total number of each type of atom is conserved, and thus the mass does not change (MS-PS1-5) **PS2.B: Types of interactions** Gravitational forces are always attractive. There is a gravitational force between any two masses, but it is very small except when one or both of the objects have large mass—e.g., earth and the sun (MS-PS2–4) Forces that act at a distance (electric, magnetic, and gravitational) can be explained by fields that extend through space and can be mapped by their effect on a test object (a charged object, or a ball, respectively) (MS-PS2–5)	**Cause and effect** Cause and effect relationships may be used to predict phenomena in natural or designed systems (MS-PS2-3)(MS-PS2-5) **Systems and system models** Models can be used to represent systems and their interactions—such as inputs, processes, and outputs—and energy and matter flows within systems (MS-PS2-1, MS-PS2-4) **Patterns** Different patterns may be observed at each of the scales at which a system is studied and can provide evidence for causality in explanations of phenomena (HS-PS1-1, HS-PS1-2, HS-PS1-3, HS-PS1-5) Different patterns may be observed at each of the scales at which a system is studied and can provide evidence for causality in explanations of phenomena (HS-PS1-1, HS-PS1-2, HS-PS1-3, HS-PS1-5)

HS.PS3.A (HS-PS2-4),(HS-PS2-5); **HS.PS3.C** (HS-PS2-1); **HS.PS4.B** (HS-PS2-5); **HS.ESS1.A** (HS-PS2-1),(HS-PS2-2),(HS-PS2-4); **HS.ESS1.B** (HS-PS2-4); **HS.ESS1.C** (HS-PS2-1),(HS-PS2-2),(HS-PS2-4); **HS.ESS2.A** (HS-PS2-5); **HS.ESS2.C** (HS-PS2-1),(HS-PS2-4); **HS.ESS3.A** (HS-PS2-4),(HS-PS2-5)
MS.PS1.A (HS-PS2-6); **MS.PS2.A** (HS-PS2-1),(HS-PS2-2),(HS-PS2-3); **MS.PS2.B** (HS-PS2-4),(HS-PS2-5),(HS-PS2-6); **MS.PS3.C** (HS-PS2-1),(HS-PS2-2),(HS-PS2-3); **MS.ESS1.B** (HS-PS2-4),(HS-PS2-5)

RST.11-12.1	Cite specific textual evidence to support analysis of science and technical texts, attending to important distinctions the author makes and to any gaps or inconsistencies in the account
RST.11-12.7	Integrate and evaluate multiple sources of information presented in diverse formats and media (e.g., quantitative data, video, multimedia) in order to address a question or solve a problem (HS-PS2-1)
WHST.9-12.2	Write informative/explanatory texts, including the narration of historical events, scientific procedures/ experiments, or technical processes
WHST.9-12.7	Conduct short as well as more sustained research projects to answer a question (including a self-generated question) or solve a problem; narrow or broaden the inquiry when appropriate; synthesize multiple sources on the subject, demonstrating understanding of the subject under investigation (HS-PS2-3, HS-PS2-5)
WHST.11-12.8	Gather relevant information from multiple authoritative print and digital sources, using advanced searches effectively; assess the strengths and limitations of each source in terms of the specific task, purpose, and audience; integrate information into the text selectively to maintain the flow of ideas, avoiding plagiarism and overreliance on any one source and following a standard format for citation
WHST.9-12.9	Draw evidence from informational texts to support analysis, reflection, and research
MP.2	Reason abstractly and quantitatively (HS-PS2-1, HS-PS2-2, HS-PS2-4)
MP.4	Model with mathematics (HS-PS2-1, HS-PS2-2, HS-PS2-4)
HSN-Q.A.1	Use units as a way to understand problems and to guide the solution of multi-step problems; choose and interpret units consistently in formulas; choose and interpret the scale and the origin in graphs and data displays (HS-PS2-1, HS-PS2-2, HS-PS2-4)
HSN-Q.A.2	Define appropriate quantities for the purpose of descriptive modeling
HSN-Q.A.3	Choose a level of accuracy appropriate to limitations on measurement when reporting quantities (HS-PS2-1, HS-PS2-2, HS-PS2-4)
HSA-SSE.A.1	Interpret expressions that represent a quantity in terms of its context (HS-PS2-1, HS-PS2-4)
HSA-SSE.B.3	Choose and produce an equivalent form of an expression to reveal and explain properties of the quantity represented by the expression (HS-PS2-1, HS-PS2-4)
HSA-CED.A.1	Create equations and inequalities in one variable and use them to solve problems (HS-PS2-1, HS-PS2-2)
HSA-CED.A.2	Create equations in two or more variables to represent relationships between quantities; graph equations on coordinate axes with labels and scales
HSA-CED.A.4	Rearrange formulas to highlight a quantity of interest, using the same reasoning as in solving equations
HSF-IF.C.7	Graph functions expressed symbolically and show key features of the graph, by in hand in simple cases and using technology for more complicated cases
HSS-ID.A.1	Represent data with plots on the real number line (dot plots, histograms, and box plots)

[93]Common Core Standards

CCSS.ELA-LITERACY.RST.11-12.5
Analyze how the text structures information or ideas into categories or hierarchies, demonstrating understanding of the information or ideas.

CCSS.ELA-LITERACY.RST.11-12.7
Integrate and evaluate multiple sources of information presented in diverse formats and media (e.g., quantitative data, video, multimedia) in order to address a question or solve a problem.

CCSS.ELA-LITERACY.RST.6-8.10
By the end of grade 8, read and comprehend science/technical texts in the grades 6–8 text complexity band independently and proficiently.

Chapter 7

[92]Standards

Standards met in the chapter based on **NGSS (New Generation Science Standards)** are:

HS-PS3-1. Create a computational model to calculate the change in the energy of one component in a system when the change in energy of the other component(s) and energy flows in and out of the system are known. [Clarification Statement: Emphasis is on explaining the meaning of mathematical expressions used in the model.] [Assessment Boundary: Assessment is limited to basic algebraic expressions or computations; to systems of two or three components; and to thermal energy, kinetic energy, and/or the energies in gravitational, magnetic, or electric fields.]

HS-PS3-2. Develop and use models to illustrate that energy at the macroscopic scale can be accounted for as a combination of energy associated with the motions of particles (objects) and energy associated with the relative position of particles (objects). [Clarification Statement: Examples of phenomena at the macroscopic scale could include the conversion of kinetic energy to thermal energy, the energy stored due to position of an object above the earth, and the energy stored between two electrically-charged plates. Examples of models could include diagrams, drawings, descriptions, and computer simulations.]

HS-PS3-3. Design, build, and refine a device that works within given constraints to convert one form of energy into another form of energy.* [Clarification Statement: Emphasis is on both qualitative and quantitative evaluations of devices. Examples of devices could include Rube Goldberg devices, wind turbines, solar cells, solar ovens, and generators. Examples of constraints could include use of renewable energy forms and efficiency.] [Assessment

Boundary: Assessment for quantitative evaluations is limited to total output for a given input. Assessment is limited to devices constructed with materials provided to students.]

HS-PS3-4. Plan and conduct an investigation to provide evidence that the transfer of thermal energy when two components of different temperature are combined within a closed system results in a more uniform energy distribution among the components in the system (second law of thermodynamics). [Clarification Statement: Emphasis is on analyzing data from student investigations and using mathematical thinking to describe the energy changes both quantitatively and conceptually. Examples of investigations could include mixing liquids at different initial temperatures or adding objects at different temperatures to water.] [Assessment Boundary: Assessment is limited to investigations based on materials and tools provided to students.]

HS-PS3-5. Develop and use a model of two objects interacting through electric or magnetic fields to illustrate the forces between objects and the changes in energy of the objects due to the interaction. [Clarification Statement: Examples of models could include drawings, diagrams, and texts, such as drawings of what happens when two charges of opposite polarity are near each other.] [Assessment Boundary: Assessment is limited to systems containing two objects.]

Science and engineering practices	Disciplinary core ideas	Crosscutting concepts
Modeling in 9–12 builds on K–8 and progresses to using, synthesizing, and developing models to predict and show relationships among variables between systems and their components in the natural and designed worlds. Develop and use a model based on evidence to illustrate the relationships between systems or between components of a system **(HS-PS3-2, HSPS3-5) planning and carrying out investigations** Planning and carrying out investigations to answer questions or test solutions to problems in 9–12 builds on K–8 experiences and progresses to include investigations that provide evidence for and test conceptual, mathematical, physical, and empirical models. Plan and conduct an investigation individually and collaboratively to produce data to serve as the basis for evidence, and in the design: Decide on types, how much, and accuracy of data needed to produce reliable measurements and consider limitations on the precision of the data (e.g., number of trials, cost, risk, time), and refine the design accordingly **(HS-PS3-4) using mathematics and computational thinking** Mathematical and computational thinking at the 9–12 level builds on K–8 and progresses to using algebraic thinking and analysis, a range of linear and nonlinear functions including trigonometric functions, exponentials and logarithms, and computational tools for statistical analysis to analyze, represent, and model data. Simple computational simulations are created and used based on mathematical models of basic assumptions. Create a computational model or simulation of a phenomenon, designed device, process, or system **(HS-PS3-1) constructing explanations and designing solutions** constructing explanations and designing solutions in 9–12 builds on K–8 experiences and progresses to explanations and designs that are supported by multiple and independent student-generated sources of evidence consistent with scientific ideas, principles, and theories. Design, evaluate, and/or refine a solution to a complex real-world problem, based on scientific knowledge, student-generated sources of evidence, prioritized criteria, and tradeoff considerations	**PS3.A: Definitions of energy** Energy is a quantitative property of a system that depends on the motion and interactions of matter and radiation within that system. That there is a single quantity called energy is due to the fact that a system's total energy is conserved, even as, within the system, energy is continually transferred from one object to another and between its various possible forms (HSPS3-1, HS-PS3-2) At the macroscopic scale, energy manifests itself in multiple ways, such as in motion, sound, light, and thermal energy (HSPS3-2, HS-PS3-3) These relationships are better understood at the microscopic scale, at which all of the different manifestations of energy can be modeled as a combination of energy associated with the motion of particles and energy associated with the configuration (relative position of the particles). In some cases the relative position energy can be thought of as stored in fields (which mediate interactions between particles). This last concept includes radiation, a phenomenon in which energy stored in fields moves across space (HS-PS3-2) **PS3.B: Conservation of energy and energy transfer** conservation of energy means that the total change of energy in any system is always equal to the total energy transferred into or out of the system (HS-PS3-1) Energy cannot be created or destroyed, but it can be transported from one place to another and transferred between systems (HS-PS3-1, HS-PS3-4) Mathematical expressions, which quantify how the stored energy in a system depends on its configuration (e.g., relative positions of charged particles, compression of a spring) and how kinetic energy depends on mass and speed, allow the concept of conservation of energy to be used to predict, and describe system behavior (HS-PS3-1) The availability of energy limits what can occur in any system (HS-PS3-1) Uncontrolled systems always evolve toward more stable states—that is, toward more uniform energy PS3.D: **Energy in chemical processes** Although energy cannot be destroyed, it can be converted to less useful forms—for example, to thermal energy in the crosscutting concepts **cause and effect** Cause and effect relationships can be suggested and predicted for complex natural and human designed systems by examining what is known about smaller scale mechanisms within the system (HS-PS3-5) **Systems and system models** When investigating or describing a system, the boundaries and initial conditions of the system need to be defined and their inputs and outputs analyzed and described using models (HS-PS3-4)	**Cause and effect** Cause and effect relationships can be suggested and predicted for complex natural and human designed systems by examining what is known about smaller scale mechanisms within the system (HS-PS3-5) **Systems and system models** when investigating or describing a system, the boundaries and initial conditions of the system need to be defined and their inputs and outputs analyzed and described using models (HS-PS3-4) Models can be used to predict the behavior of a system, but these predictions have limited precision and reliability due to the assumptions and approximations inherent in models (HSPS3-1) **Energy and matter** Changes of energy and matter in a system can be described in terms of energy and matter flows into, out of, and within that system (HSPS3-3) Energy cannot be created or destroyed—only moves between one place and another place, between objects and/or fields, or between systems (HS-PS3–2)

RST.11-12.1 Cite specific textual evidence to support analysis of science and technical texts, attending to important distinctions the author makes and to any gaps or inconsistencies in the account (HS-PS3-4).
WHST.9-12.7 Conduct short as well as more sustained research projects to answer a question (including a self-generated question) or solve a problem; narrow or broaden the inquiry when appropriate; synthesize multiple sources on the subject, demonstrating understanding of the subject under investigation (HS-PS3-3, HS-PS3-4, HS-PS3-5).
WHST.11-12.8 Gather relevant information from multiple authoritative print and digital sources, using advanced searches effectively; assess the strengths and limitations of each source in terms of the specific task, purpose, and audience; integrate information into the text selectively to maintain the flow of ideas, avoiding plagiarism and overreliance on any one source and following a standard format for citation (HS-PS3-4, HS-PS3-5).
WHST.9-12.9 Draw evidence from informational texts to support analysis, reflection, and research (HS-PS3-4, HS-PS3-5).
SL.11-12.5 Make strategic use of digital media (e.g., textual, graphical, audio, visual, and interactive elements) in presentations to enhance understanding of findings, reasoning, and evidence and to add interest (HS-PS3-1, HS-PS3-2, HS-PS3-5).
Mathematics:
MP.2 Reason abstractly and quantitatively (HS-PS3-1, HS-PS3-2, HS-PS3-3, HS-PS3-4, HS-PS3-5).
MP.4 Model with mathematics (HS-PS3-1, HS-PS3-2, HS-PS3-3, HS-PS3-4, HS-PS3-5).
HSN-Q.A.1 Use units as a way to understand problems and to guide the solution of multi-step problems; choose and interpret units consistently in formulas; choose and interpret the scale and the origin in graphs and data displays (HS-PS3-1, HS-PS3-3).
HSN-Q.A.2 Define appropriate quantities for the purpose of descriptive modeling (HS-PS3-1, HS-PS3-3).
HSN-Q.A.3 Choose a level of accuracy appropriate to limitations on measurement when reporting quantities (HS-PS3-1, HS-PS3-3).

[93]Common Core Standards

CCSS.ELA-LITERACY.RST.11-12.5
Analyze how the text structures information or ideas into categories or hierarchies, demonstrating understanding of the information or ideas.

CCSS.ELA-LITERACY.RST.11-12.7
Integrate and evaluate multiple sources of information presented in diverse formats and media (e.g., quantitative data, video, multimedia) in order to address a question or solve a problem.

CCSS.ELA-LITERACY.RST.6-8.10
By the end of grade 8, read and comprehend science/technical texts in the grades 6–8 text complexity band independently and proficiently.

Chapter 8

[92]Standards

Standards met in the chapter based on **NGSS (New Generation Science Standards)** are:

MS-PS2-1. Apply Newton's Third Law to design a solution to a problem involving the motion of two colliding objects.* [Clarification Statement: Examples of practical problems could include the impact of collisions between two cars, between a car and stationary objects, and between a meteor and a space vehicle.] [Assessment Boundary: Assessment is limited to vertical or horizontal interactions in one dimension.]

MS-PS2-2. Plan an investigation to provide evidence that the change in an object's motion depends on the sum of the forces on the object and the mass of the object. [Clarification Statement: Emphasis is on balanced (Newton's First Law) and unbalanced forces in a system, qualitative comparisons of forces, mass and changes in motion (Newton's Second Law), frame of reference, and specification of units.] [Assessment Boundary: Assessment is limited to forces and changes in motion in one dimension in an inertial reference frame and to change in one variable at a time. Assessment does not include the use of trigonometry.]

MS-PS2-3. Ask questions about data to determine the factors that affect the strength of electric and magnetic forces. [Clarification Statement: Examples of devices that use electric and magnetic forces could include electromagnets, electric motors, or generators. Examples of data could include the effect of the number of turns of wire on the strength of an electromagnet, or the effect of increasing the number or strength of magnets on the speed of an electric motor.] [Assessment Boundary: Assessment about questions that require quantitative answers is limited to proportional reasoning and algebraic thinking.]

HS-PS1-1. Use the periodic table as a model to predict the relative properties of elements based on the patterns of electrons in the outermost energy level of atoms. [Clarification Statement: Examples of properties that could be predicted from patterns could include reactivity of metals, types of bonds formed, numbers of bonds formed, and reactions with oxygen.] [Assessment Boundary: Assessment is limited to main group elements. Assessment does not include quantitative understanding of ionization energy beyond relative trends.] [Assessment Boundary: techniques. Assessment does not include complex chemical reactions.]

HS-PS1-3. Plan and conduct an investigation to gather evidence to compare the structure of substances at the bulk scale to infer the strength of electrical forces between particles. [Clarification Statement: Emphasis is on understanding the strengths of forces between particles, not on naming specific intermolecular forces (such as dipole-dipole). Examples of particles could include ions, atoms, molecules, and networked materials (such as graphite). Examples of bulk properties of substances could include the melting point and boiling point, vapor pressure, and surface tension.] [Assessment Boundary: Assessment does not include Raoult's law calculations of vapor pressure.]

HS-PS1-4. Develop a model to illustrate that the release or absorption of energy from a chemical reaction system depends upon the changes in total bond energy. [Clarification Statement: Emphasis is on the idea that a chemical reaction is a system that affects the energy change. Examples of models could include molecular-level drawings and diagrams of reactions, graphs showing the relative energies of reactants and products, and representations showing energy is conserved.] [Assessment Boundary: Assessment does not include calculating the total bond energy changes during a chemical reaction from the bond energies of reactants and products.]

HS-PS1-5. Apply scientific principles and evidence to provide an explanation about the effects of changing the temperature or concentration of the reacting particles on the rate at which a reaction occurs. [Clarification Statement: Emphasis is on student reasoning that focuses on the number and energy of collisions between molecules.] [Assessment Boundary: Assessment is limited to simple reactions in which there are only two reactants; evidence from temperature, concentration, and rate data; and qualitative relationships between rate and temperature.]

Science and engineering practices	Disciplinary core ideas	Crosscutting concepts
Constructing explanations and designing solutions Constructing explanations and designing solutions in 6–8 builds on K–5 experiences and progresses to include constructing explanations and designing solutions supported by multiple sources of evidence consistent with scientific ideas, principles, and theories Apply scientific ideas or principles to design an object, tool, process or system (MS-PS2-1) **Obtaining, evaluating, and communicating information** Obtaining, evaluating, and communicating information in 6–8 builds on K–5 and progresses to evaluating the merit and validity of ideas and methods Gather, read, and synthesize information from multiple appropriate sources and assess the credibility, accuracy, and possible bias of each publication and methods used, and describe how they are supported or not supported by evidence (MS-PS1-3) Modeling in 9–12 builds on K–8 and progresses to using, synthesizing, and developing models to predict and show relationships among variables between systems and their components in the natural and designed worlds Use a model to predict the relationships between systems or between components of a system (HS-PS1-1) **HS-PS1-1.** Use the periodic table as a model to predict the relative properties of elements based on the patterns of electrons in the outermost energy level of atoms. [Clarification statement: Examples of properties that could be predicted from patterns could include reactivity of metals, types of bonds formed, numbers of bonds formed, and reactions with oxygen.] [assessment boundary: Assessment is limited to main group elements. Assessment does not include quantitative understanding of ionization energy beyond relative trends.] **HS-PS1-3. Planning and conducting investigations** Plan and conduct an investigation to gather evidence to compare the structure of substances at the bulk scale to infer the strength of electrical forces between particles. [Clarification statement: Emphasis is on understanding the strengths of forces between particles, not on naming specific intermolecular forces (such as dipole-dipole) Examples of particles could include ions, atoms, molecules, and networked materials (such as graphite). Examples of bulk properties of substances could include the melting point and boiling point, vapor pressure, and surface tension.] [Assessment boundary: Assessment does not include Raoult's law calculations of vapor pressure.]	**PS1.A: Structure and properties of matter** Each pure substance has characteristic physical and chemical properties (for any bulk quantity under given conditions) that can be used to identify it (MS-PS1-2, MS-PS1-3) Each atom has a charged substructure consisting of a nucleus, which is made of protons and neutrons, surrounded by electrons (HS-PS1-1) **PS1.B: Chemical reactions** Substances react chemically in characteristic ways. In a chemical process, the atoms that make up the original substances are regrouped into different molecules, and these new substances have different properties from those of the reactants (MS-PS1-2, MS-PS1-3, MS-PS1-5) The total number of each type of atom is conserved, and thus the mass does not change (MS-PS1-5) **PS2.B: Types of interactions** Forces that act at a distance (electric, magnetic, and gravitational) can be explained by fields that extend through space and can be mapped by their effect on a test object (a charged object, or a ball, respectively) (MS-PS2–5)	**Cause and effect** Cause and effect relationships may be used to predict phenomena in natural or designed system. (MS-PS2-3),(MS-PS2-5) **Systems and system models** Models can be used to represent systems and their interactions—Such as inputs, processes, and outputs—And energy and matter flows within systems (MS-PS2-1, MS-PS2-4) **Patterns** Different patterns may be observed at each of the scales at which a system is studied and can provide evidence for causality in explanations of phenomena (HS-PS1-1, HS-PS1-2, HS-PS1-3, HS- PS1-5)

RST.11-12.1	Cite specific textual evidence to support analysis of science and technical texts, attending to important distinctions the author makes and to any gaps or inconsistencies in the account
RST.11-12.7	Integrate and evaluate multiple sources of information presented in diverse formats and media (e.g., quantitative data, video, multimedia) in order to address a question or solve a problem (HS-PS2-1)
WHST.9-12.2	Write informative/explanatory texts, including the narration of historical events, scientific procedures/experiments, or technical processes
WHST.9-12.7	Conduct short as well as more sustained research projects to answer a question (including a self-generated question) or solve a problem; narrow or broaden the inquiry when appropriate; synthesize multiple sources on the subject, demonstrating understanding of the subject under investigation (HS-PS2-3, HS-PS2-5)
WHST.11-12.8	Gather relevant information from multiple authoritative print and digital sources, using advanced searches effectively; assess the strengths and limitations of each source in terms of the specific task, purpose, and audience; integrate information into the text selectively to maintain the flow of ideas, avoiding plagiarism and overreliance on any one source and following a standard format for citation
WHST.9-12.9	Draw evidence from informational texts to support analysis, reflection, and research
MP.2	Reason abstractly and quantitatively (HS-PS2-1, HS-PS2-2, HS-PS2-4)
MP.4	Model with mathematics (HS-PS2-1, HS-PS2-2, HS-PS2-4)
HSN-Q.A.1	Use units as a way to understand problems and to guide the solution of multi-step problems; choose and interpret units consistently in formulas; choose and interpret the scale and the origin in graphs and data displays (HS-PS2-1, HS-PS2-2, HS-PS2-4)
HSN-Q.A.2	Define appropriate quantities for the purpose of descriptive modeling
HSN-Q.A.3	Choose a level of accuracy appropriate to limitations on measurement when reporting quantities (HS-PS2-1, HS-PS2-2, HS-PS2-4)
HSA-SSE.A.1	Interpret expressions that represent a quantity in terms of its context (HS-PS2-1, HS-PS2-4)
HSA-SSE.B.3	Choose and produce an equivalent form of an expression to reveal and explain properties of the quantity represented by the expression (HS-PS2-1, HS-PS2-4)
HSA-CED.A.1	Create equations and inequalities in one variable and use them to solve problems (HS-PS2-1, HS-PS2-2)
HSA-CED.A.2	Create equations in two or more variables to represent relationships between quantities; graph equations on coordinate axes with labels and scales
HSA-CED.A.4	Rearrange formulas to highlight a quantity of interest, using the same reasoning as in solving equations
HSF-IF.C.7	Graph functions expressed symbolically and show key features of the graph, by in hand in simple cases and using technology for more complicated cases
HSS-ID.A.1	Represent data with plots on the real number line (dot plots, histograms, and box plots).

[93]Common Core Standards

CCSS.ELA-LITERACY.RST.11-12.4
Determine the meaning of symbols, key terms, and other domain-specific words and phrases as they are used in a specific scientific or technical context relevant to *grades 11–12 texts and topics.*

CCSS.ELA-LITERACY.RST.11-12.7
Integrate and evaluate multiple sources of information presented in diverse formats and media (e.g., quantitative data, video, multimedia) in order to address a question or solve a problem.

CCSS.ELA-LITERACY.RST.11-12.10
By the end of grade 12, read and comprehend science/technical texts in the grades 11-CCR text complexity band independently and proficiently grades 6–8 text complexity band independently and proficiently

CCSS.ELA-LITERACY.RST.6-8.4
Determine the meaning of symbols, key terms, and other domain-specific words and phrases as they are used in a specific scientific or technical context relevant to grades 6–8 texts and topics.

Chapter 9

[92]Standards

Standards met in the chapter based on **NGSS (New Generation Science Standards)** are:

MS-PS2-1. Apply Newton's Third Law to design a solution to a problem involving the motion of two colliding objects.* [Clarification Statement: Examples of practical problems could include the impact of collisions between two cars, between a car and stationary objects, and between a meteor and a space vehicle.] [Assessment Boundary: Assessment is limited to vertical or horizontal interactions in one dimension.]

MS-PS2-2. Plan an investigation to provide evidence that the change in an object's motion depends on the sum of the forces on the object and the mass of the object. [Clarification Statement: Emphasis is on balanced (Newton's First Law) and unbalanced forces in a system, qualitative comparisons of forces, mass, and changes in motion (Newton's Second Law), frame of reference, and specification of units.] [Assessment Boundary: Assessment is limited to forces and changes in motion in one dimension in an inertial reference frame and to change in one variable at a time. Assessment does not include the use of trigonometry.]

MS-PS2-3. Ask questions about data to determine the factors that affect the

strength of electric and magnetic forces. [Clarification Statement: Examples of devices that use electric and magnetic forces could include electromagnets, electric motors, or generators. Examples of data could include the effect of the number of turns of wire on the strength of an electromagnet, or the effect of increasing the number or strength of magnets on the speed of an electric motor.] [Assessment Boundary: Assessment about questions that require quantitative answers is limited to proportional reasoning and algebraic thinking.]

HS-PS1-1. Use the periodic table as a model to predict the relative properties of elements based on the patterns of electrons in the outermost energy level of atoms. [Clarification Statement: Examples of properties that could be predicted from patterns could include reactivity of metals, types of bonds formed, numbers of bonds formed, and reactions with oxygen.] [Assessment Boundary: Assessment is limited to main group elements. Assessment does not include quantitative understanding of ionization energy beyond relative trends.] [Assessment Boundary: techniques. Assessment does not include complex chemical reactions.]

HS-PS1-3. Plan and conduct an investigation to gather evidence to compare the structure of substances at the bulk scale to infer the strength of electrical forces between particles. [Clarification Statement: Emphasis is on understanding the strengths of forces between particles, not on naming specific intermolecular forces (such as dipole-dipole). Examples of particles could include ions, atoms, molecules, and networked materials (such as graphite). Examples of bulk properties of substances could include the melting point and boiling point, vapor pressure, and surface tension.] [Assessment Boundary: Assessment does not include Raoult's law calculations of vapor pressure.]

HS-PS1-4. Develop a model to illustrate that the release or absorption of energy from a chemical reaction system depends upon the changes in total bond energy. [Clarification Statement: Emphasis is on the idea that a chemical reaction is a system that affects the energy change. Examples of models could include molecular-level drawings and diagrams of reactions, graphs showing the relative energies of reactants and products, and representations showing energy is conserved.] [Assessment Boundary: Assessment does not include calculating the total bond energy changes during a chemical reaction from the bond energies of reactants and products.]

HS-PS1-5. Apply scientific principles and evidence to provide an explanation about the effects of changing the temperature or concentration of the reacting particles on the rate at which a reaction occurs. [Clarification Statement: Emphasis is on student reasoning that focuses on the number and energy of collisions between molecules.] [Assessment Boundary: Assessment is limited to simple reactions in which there are only two reactants; evidence from temperature, concentration, and rate data; and qualitative relationships between rate and temperature.]

Science and engineering practices	Disciplinary core ideas	Crosscutting concepts
Constructing explanations and designing solutions Constructing explanations and designing solutions in 6–8 builds on K–5 experiences and progresses to include constructing explanations and designing solutions supported by multiple sources of evidence consistent with scientific ideas, principles, and theories Apply scientific ideas or principles to design an object, tool, process or system (MS-PS2-1) **Obtaining, evaluating, and communicating information** Obtaining, evaluating, and communicating information in 6–8 builds on K–5 and progresses to evaluating the merit and validity of ideas and methods Gather, read, and synthesize information from multiple appropriate sources and assess the credibility, accuracy, and possible bias of each publication and methods used, and describe how they are supported or not supported by evidence (MS-PS1-3) Modeling in 9–12 builds on K–8 and progresses to using, synthesizing, and developing models to predict and show relationships among variables between systems and their components in the natural and designed worlds. Use a model to predict the relationships between systems or between components of a system (HS-PS1-1) **HS-PS1-1.** Use the periodic table as a model to predict the relative properties of elements based on the patterns of electrons in the outermost energy level of atoms. [Clarification statement: Examples of properties that could be predicted from patterns could include reactivity of metals, types of bonds formed, numbers of bonds formed, and reactions with oxygen.] [Assessment boundary: Assessment is limited to main group elements. Assessment does not include quantitative understanding of ionization energy beyond relative trends.] **HS-PS1-3. Planning and conducting investigations** Plan and conduct an investigation to gather evidence to compare the structure of substances at the bulk scale to infer the strength of electrical forces between particles. [Clarification statement: Emphasis is on understanding the strengths of forces between particles, not on naming specific intermolecular forces (such as dipole-dipole). Examples of particles could include ions, atoms, molecules, and networked materials (such as graphite). Examples of bulk properties of substances could include the melting point and boiling point, vapor pressure, and surface tension.] [Assessment boundary: Assessment does not include Raoult's law calculations of vapor pressure.]	**PS1.A: Structure and properties of matter** Each pure substance has characteristic physical and chemical properties (for any bulk quantity under given conditions) that can be used to identify it (MS-PS1-2, MS-PS1-3) Each atom has a charged substructure consisting of a nucleus, which is made of protons and neutrons, surrounded by electrons (HS-PS1-1) **PS1.B: Chemical reactions** Substances react chemically in characteristic ways. In a chemical process, the atoms that make up the original substances are regrouped into different molecules, and these new substances have different properties from those of the reactants (MS-PS1-2, MS-PS1-3, MS-PS1-5) The total number of each type of atom is conserved, and thus the mass does not change (MS-PS1-5) **PS2.B: Types of interactions** Forces that act at a distance (electric, magnetic, and gravitational) can be explained by fields that extend through space and can be mapped by their effect on a test object (a charged object, or a ball, respectively) (MS-PS2–5)	**Cause and effect** Cause and effect relationships may be used to predict phenomena in natural or designed systems (MS-PS2-3, MS-PS2-5) **Systems and system models** Models can be used to represent systems and their interactions—such as inputs, processes, and outputs—and energy and matter flows within systems (MS-PS2-1, MS-PS2-4) **Patterns** Different patterns may be observed at each of the scales at which a system is studied and can provide evidence for causality in explanations of phenomena (HS-PS1-1, HS-PS1-2, HS-PS1-3, HS-PS1-5)

HS.PS3.A (HS-PS2-4),(HS-PS2-5); **HS.PS3.C** (HS-PS2-1); **HS.PS4.B** (HS-PS2-5); **HS. ESS1.A** (HS-PS2-1),(HS-PS2-2),(HS-PS2-4);
HS.ESS1.B (HS-PS2-4); **HS.ESS1.C** (HS-PS2-1),(HS-PS2-2),(HS-PS2-4); **HS.ESS2.A** (HS-PS2-5); **HS.ESS2.C** (HS-PS2-1),(HS-PS2-4); **HS.ESS3.A** (HS-PS2-4),(HS-PS2-5)

MS.PS1.A (HS-PS2-6); **MS.PS2.A** (HS-PS2-1),(HS-PS2-2),(HS-PS2-3); **MS.PS2.B** (HS-PS2-4),(HS-PS2-5),(HS-PS2-6); **MS.PS3.C** (HS-PS2-1),(HS-PS2-2),(HS-PS2-3); **MS.ESS1.B** (HS-PS2-4),(HS-PS2-5)

RST.11-12.1	Cite specific textual evidence to support analysis of science and technical texts, attending to important distinctions the author makes and to any gaps or inconsistencies in the account
RST.11-12.7	Integrate and evaluate multiple sources of information presented in diverse formats and media (e.g., quantitative data, video, multimedia) in order to address a question or solve a problem (HS-PS2-1)
WHST.9-12.2	Write informative/explanatory texts, including the narration of historical events, scientific procedures/experiments, or technical processes
WHST.9-12.7	Conduct short as well as more sustained research projects to answer a question (including a self-generated question) or solve a problem; narrow or broaden the inquiry when appropriate; synthesize multiple sources on the subject, demonstrating understanding of the subject under investigation (HS-PS2-3, HS-PS2-5)
WHST.11-12.8	Gather relevant information from multiple authoritative print and digital sources, using advanced searches effectively; assess the strengths and limitations of each source in terms of the specific task, purpose, and audience; integrate information into the text selectively to maintain the flow of ideas, avoiding plagiarism and overreliance on any one source and following a standard format for citation
WHST.9-12.9	Draw evidence from informational texts to support analysis, reflection, and research
MP.2	Reason abstractly and quantitatively (HS-PS2-1, HS-PS2-2, HS-PS2-4)
MP.4	Model with mathematics (HS-PS2-1, HS-PS2-2, HS-PS2-4)
HSN-Q.A.1	Use units as a way to understand problems and to guide the solution of multi-step problems; choose and interpret units consistently in formulas; choose and interpret the scale and the origin in graphs and data displays (HS-PS2-1, HS-PS2-2, HS-PS2-4)
HSN-Q.A.2	Define appropriate quantities for the purpose of descriptive modeling
HSN-Q.A.3	Choose a level of accuracy appropriate to limitations on measurement when reporting quantities (HS-PS2-1, HS-PS2-2, HS-PS2-4)
HSA-SSE.A.1	Interpret expressions that represent a quantity in terms of its context (HS-PS2-1, HS-PS2-4)
HSA-SSE.B.3	Choose and produce an equivalent form of an expression to reveal and explain properties of the quantity represented by the expression (HS-PS2-1, HS-PS2-4)
HSA-CED.A.1	Create equations and inequalities in one variable and use them to solve problems (HS-PS2-1, HS-PS2-2)
HSA-CED.A.2	Create equations in two or more variables to represent relationships between quantities; graph equations on coordinate axes with labels and scales
HSA-CED.A.4	Rearrange formulas to highlight a quantity of interest, using the same reasoning as in solving equations
HSF-IF.C.7	Graph functions expressed symbolically and show key features of the graph, by in hand in simple cases and using technology for more complicated cases
HSS-ID.A.1	Represent data with plots on the real number line (dot plots, histograms, and box plots)

[93]Common Core Standards

CCSS.ELA-LITERACY.RST.11-12.4
Determine the meaning of symbols, key terms, and other domain-specific words and phrases as they are used in a specific scientific or technical context relevant to *grades 11–12 texts and topics.*

CCSS.ELA-LITERACY.RST.11-12.7
Integrate and evaluate multiple sources of information presented in diverse formats and media (e.g., quantitative data, video, multimedia) in order to address a question or solve a problem.

CCSS.ELA-LITERACY.RST.11-12.10
By the end of grade 12, read and comprehend science/technical texts in the grades 11-CCR text complexity band independently and proficiently.
grades 6–8 text complexity band independently and proficiently

CCSS.ELA-LITERACY.RST.6-8.4
Determine the meaning of symbols, key terms, and other domain-specific words and phrases as they are used in a specific scientific or technical context relevant to grades 6–8 texts and topics.

Chapter 10

[92]Standards

Standards met in the chapter based on **NGSS (New Generation Science Standards**) are:

MS-PS1-4. Develop a model that predicts and describes changes in particle motion, temperature, and state of a pure substance when thermal energy is added or removed. [Clarification Statement: Emphasis is on qualitative molecular-level models of solids, liquids, and gases to show that adding or removing thermal energy increases or decreases kinetic energy of the particles until a change of state occurs. Examples of models could include drawings and diagrams. Examples of particles could include molecules or inert atoms. Examples of pure substances could include water, carbon dioxide, and helium.]

MS-PS2-3. Ask questions about data to determine the factors that affect the strength of electric and magnetic forces. [Clarification Statement: Examples of devices that use electric and magnetic forces could include electromagnets, electric motors, or generators. Examples of data could include the effect of the number of turns of wire on the strength of an electromagnet, or the effect of increasing the number or strength of magnets on the speed of an electric motor.] [Assessment Boundary: Assessment about questions that require quantitative answers is limited to proportional reasoning and algebraic thinking.]

HS-PS1-5. Apply scientific principles and evidence to provide an explanation about the effects of changing the temperature or concentration of the reacting particles on the rate at which a reaction occurs. [Clarification Statement: Emphasis is on student reasoning that focuses on the number and energy of collisions between molecules.] [Assessment Boundary: Assessment is limited to simple reactions in which there are only two reactants; evidence from temperature, concentration, and rate data; and qualitative relationships between rate and temperature.]

HS-PS1-3. Plan and conduct an investigation to gather evidence to compare the structure of substances at the bulk scale to infer the strength of electrical forces between particles. [Clarification Statement: Emphasis is on understanding the strengths of forces between particles, not on naming specific intermolecular forces (such as dipole-dipole). Examples of particles could include ions, atoms, molecules, and networked materials (such as graphite). Examples of bulk properties of substances could include the melting point and boiling point, vapor pressure, and surface tension.] [Assessment Boundary: Assessment does not include Raoult's law calculations of vapor pressure.]

HS-PS1-5. Apply scientific principles and evidence to provide an explanation about the effects of changing the temperature or concentration of the reacting particles on the rate at which a reaction occurs. [Clarification Statement: Emphasis is on student reasoning that focuses on the number and energy of collisions between molecules.] [Assessment Boundary: Assessment is limited to simple reactions in which there are only two reactants; evidence from temperature, concentration, and rate data; and qualitative relationships between rate and temperature.]

Science and engineering practices	Disciplinary core ideas	Crosscutting concepts
Asking questions and defining problems Asking questions and defining problems in grades 6–8 builds from grades K–5 experiences and progresses to specifying relationships between variables, and clarifying arguments and models Ask questions that can be investigated within the scope of the classroom, outdoor environment, and museums and other public facilities with available resources and, when appropriate, frame a hypothesis based on observations and scientific principles (MS-PS2-3). **Planning and carrying out investigations** Planning and carrying out investigations to answer questions or test solutions to problems in 6–8 builds on K–5 experiences and progresses to include investigations that use multiple variables and provide evidence to support explanations or design solutions. Plan an investigation individually and collaboratively, and in the design: Identify independent and dependent variables and controls, what tools are needed to do the gathering, how measurements will be recorded, and how many data are needed to support a claim (MS-PS2-2). **Planning and carrying out investigations** Planning and carrying out investigations in 9–12 builds on K-8 experiences and progresses to include investigations that provide evidence for and test conceptual, mathematical, physical, and empirical models Plan and conduct an investigation individually and collaboratively to produce data to serve as the basis for evidence, and in the design: Decide on types, how much, and accuracy of data needed to produce reliable measurements and consider limitations on the precision of the data (e.g., number of trials, cost, risk, time), and refine the design accordingly (HS-PS1-3) **Constructing explanations and designing solutions** Constructing explanations and designing solutions in 6–8 builds on K–5 experiences and progresses to include constructing explanations and designing solutions supported by multiple sources of evidence consistent with scientific ideas, principles, and theories. Apply scientific ideas or principles to design an object, tool, process, or system (MS-PS2-1). **Developing and using models** Modeling in 9–12 builds on K–8 and progresses to using, synthesizing, and developing models to predict and show relationships among variables between systems and their components in the natural and designed worlds Use a model to predict the relationships between systems or between components of a system (HS-PS1-1)	**PS2.A: Forces and motion** For any pair of interacting objects, the force exerted by the first object on the second object is equal in strength to the force that the second object exerts on the first, but in the opposite direction (Newton's Third Law) (MS-PS2-1). The motion of an object is determined by the sum of the forces acting on it; if the total force on the object is not zero, its motion will change. The greater the mass of the object, the greater the force needed to achieve the same change in motion. For any given object, a larger force causes a larger change in motion (MS-PS2-2). All positions of objects and the directions of forces and motions must be described in an arbitrarily chosen reference frame and arbitrarily chosen units of size. In order to share information with other people, these choices must also be shared (MSPS2-2). **PS1.A: Structure and properties of matter** Each atom has a charged substructure consisting of a nucleus, which is made of protons and neutrons, surrounded by electrons. (HS-PS1-1) The periodic table orders elements horizontally by the number of protons in the atom's nucleus and places those with similar chemical properties in columns. The repeating patterns of this table reflect patterns of outer electron states. (HS-PS1–1),(HS-PS1–2) The structure and interactions of matter at the bulk Scale are determined by electrical forces within and between atoms. (HS-PS1–3), A stable molecule has less energy than the same set of atoms separated; one must provide at least this energy in order to take the molecule apart. (HS-PS1-4)	**Cause and effect** Cause and effect relationships may be used to predict phenomena in natural or designed systems (MS-PS2-3, MS-PS2-5). **Systems and system models** Models can be used to represent systems and their interactions—such as inputs, processes, and outputs—and energy and matter flows within systems (MS-PS2-1, MS-PS2-4). **Stability and change** Explanations of stability and change in natural or designed systems can be constructed by examining the changes over time and forces at different scales (MS-PS2-2) **Patterns** Different patterns may be observed at each of the scales at which a system is studied and can provide evidence for causality in explanations of phenomena (HS-PS1-1, HS-PS1-2, HS-PS1-3, HS-PS1-5) ***Connections to engineering, technology, and applications of science*** **Influence of science, engineering, and technology on society and the natural world** The uses of technologies and any limitations on their use are driven by individual or societal needs, desires, and values; by the findings of scientific research; and by differences in such factors as climate, natural resources, and economic conditions (MS-PS2-1)

Science and engineering practices	Disciplinary core ideas	Crosscutting concepts
MS-PS2 Motion and Stability: Forces and Interactions *The performance expectations marked with an asterisk integrate traditional science content with engineering through a Practice or Disciplinary Core Idea. The section entitled "Disciplinary Core Ideas" is reproduced verbatim from A Framework for K-12 Science Education: Practices, Cross-Cutting Concepts, and Core Ideas. Integrated and reprinted with permission from the National Academy of Sciences **RST.6-8.1** Cite specific textual evidence to support analysis of science and technical texts, attending to the precise details of explanations or descriptions (MS-PS2-1, MSPS2-3) **RST.6-8.3** Follow precisely a multistep procedure when carrying out experiments, taking measurements, or performing technical tasks (MS-PS2-1, MS-PS2-2, MS-PS2-5) **WHST.6-8.1** Write arguments focused on discipline-specific content (MS-PS2-4) **WHST.6-8.7** Conduct short research projects to answer a question (including a self-generated question), drawing on several sources and generating additional related, focused questions that allow for multiple avenues of exploration (MS-PS2-1, MS-PS2-2, MS-PS2-5) **Mathematics:** **MP.2** Reason abstractly and quantitatively (MS-PS2-1, MS-PS2-2, MS-PS2-3) **6.NS.C.5** Understand that positive and negative numbers are used together to describe quantities having opposite directions or values; use positive and negative numbers to represent quantities in real-world contexts, explaining the meaning of 0 in each situation (MS-PS2-1) **6.EE.A.2** Write, read, and evaluate expressions in which letters stand for numbers (MS-PS2-1, MS-PS2-2) **7.EE.B.3** Solve multi-step real-life and mathematical problems posed with positive and negative rational numbers in any form, using tools strategically. Apply properties of operations to calculate with numbers in any form; convert between forms as appropriate; and assess the reasonableness of answers using mental computation and estimation strategies (MS-PS2-1, MS-PS2-2) **7.EE.B.4** Use variables to represent quantities in a real-world or mathematical problem, and construct simple equations and inequalities to solve problems by reasoning about the quantities (MS-PS2-1, MS-PS2-2)		

[93]Common Core Standards

CCSS.ELA-LITERACY.RST.11-12.4
Determine the meaning of symbols, key terms, and other domain-specific words and phrases as they are used in a specific scientific or technical context relevant to *grades 11–12 texts and topics.*

CCSS.ELA-LITERACY.RST.11-12.9
Synthesize information from a range of sources (e.g., texts, experiments, simulations) into a coherent understanding of a process, phenomenon, or concept, resolving conflicting information when possible.

CCSS.ELA-LITERACY.RST.11-12.10
By the end of grade 12, read and comprehend science/technical texts in the grades 11-CCR text complexity band independently and proficiently.

CCSS.ELA-LITERACY.RST.6-8.4
Determine the meaning of symbols, key terms, and other domain-specific words and phrases as they are used in a specific scientific or technical context relevant to grades 6–8 texts and topics.

Chapter 11

[92]Standards

Standards met in the chapter based on **NGSS (New Generation Science Standards)** are:

MS-PS1-4. Develop a model that predicts and describes changes in particle motion, temperature, and state of a pure substance when thermal energy is added or removed. [Clarification Statement: Emphasis is on qualitative molecular-level models of solids, liquids, and gases to show that adding or removing thermal energy increases or decreases kinetic energy of the particles until a change of state occurs. Examples of models could include drawings and diagrams. Examples of particles could include molecules or inert atoms. Examples of pure substances could include water, carbon dioxide, and helium.]

MS-PS2-3. Ask questions about data to determine the factors that affect the strength of electric and magnetic forces. [Clarification Statement: Examples of devices that use electric and magnetic forces could include electromagnets, electric motors, or generators. Examples of data could include the effect of the number of turns of wire on the strength of an electromagnet, or the effect of increasing the number or strength of magnets on the speed of an electric motor.] [Assessment Boundary: Assessment about questions that require quantitative answers is limited to proportional reasoning and algebraic thinking.]

MS-PS3-3. Apply scientific principles to design, construct, and test a device that either minimizes or maximizes thermal energy transfer.* [Clarification Statement: Examples of devices could include an insulated box, a solar cooker, and a Styrofoam cup.] [Assessment Boundary: Assessment does not include calculating the total amount of thermal energy transferred.]

HS-PS1-5. Apply scientific principles and evidence to provide an explanation about the effects of changing the temperature or concentration of the reacting particles on the rate at which a reaction occurs. [Clarification Statement: Emphasis is on student reasoning that focuses on the number and energy of collisions between molecules.] [Assessment Boundary: Assessment is limited to simple reactions in which there are only two reactants; evidence from temperature, concentration, and rate data; and qualitative relationships between rate and temperature.]

HS-PS1-3. Plan and conduct an investigation to gather evidence to compare the structure of substances at the bulk scale to infer the strength of electrical forces between particles. [Clarification Statement: Emphasis is on understanding the strengths of forces between particles, not on naming specific intermolecular forces (such as dipole-dipole). Examples of particles could include ions, atoms, molecules, and networked materials (such as graphite). Examples of bulk properties of substances could include the melting point and boiling point, vapor pressure, and surface tension.] [Assessment Boundary: Assessment does not include Raoult's law calculations of vapor pressure.]

HS-PS1-5. Apply scientific principles and evidence to provide an explanation about the effects of changing the temperature or concentration of the reacting particles on the rate at which a reaction occurs. [Clarification Statement: Emphasis is on student reasoning that focuses on the number and energy of collisions between molecules.] [Assessment Boundary: Assessment is limited to simple reactions in which there are only two reactants; evidence from temperature, concentration, and rate data; and qualitative relationships between rate and temperature.]

Science and engineering practices	Disciplinary core ideas	Crosscutting concepts
Asking questions and defining problems Asking questions and defining problems in grades 6–8 builds from grades K–5 experiences and progresses to specifying relationships between variables, and clarifying arguments and models Ask questions that can be investigated within the scope of the classroom, outdoor environment, and museums and other public facilities with available resources and, when appropriate, frame a hypothesis based on observations and scientific principles (MS-PS2-3). **Planning and carrying out investigations** Planning and carrying out investigations to answer questions or test solutions to problems in 6–8 builds on K–5 experiences and progresses to include investigations that use multiple variables and provide evidence to support explanations or design solutions. Plan an investigation individually and collaboratively, and in the design: Identify independent and dependent variables and controls, what tools are needed to do the gathering, how measurements will be recorded, and how many data are needed to support a claim (MS-PS2-2). **Planning and carrying out investigations** Planning and carrying out investigations in 9–12 builds on K-8 experiences and progresses to include investigations that provide evidence for and test conceptual, mathematical, physical, and empirical models. Plan and conduct an investigation individually and collaboratively to produce data to serve as the basis for evidence, and in the design: Decide on types, how much, and accuracy of data needed to produce reliable measurements and consider limitations on the precision of the data (e.g., number of trials, cost, risk, time), and refine the design accordingly (HS-PS1–3) **Constructing explanations and designing solutions** Constructing explanations and designing solutions in 6–8 builds on K–5 experiences and progresses to include constructing explanations and designing solutions supported by multiple sources of evidence consistent with scientific ideas, principles, and theories Apply scientific ideas or principles to design an object, tool, process or system (MS-PS2-1). **Developing and using models** Modeling in 9–12 builds on K–8 and progresses to using, synthesizing, and developing models to predict and show relationships among variables between systems and their components in the natural and designed worlds Use a model to predict the relationships between systems or between components of a system (HS-PS1-1)	**PS2.A: Forces and motion** For any pair of interacting objects, the force exerted by the first object on the second object is equal in strength to the force that the second object exerts on the first, but in the opposite direction (Newton's Third Law) (MS-PS2-1). The motion of an object is determined by the sum of the forces acting on it; if the total force on the object is not zero, its motion will change. The greater the mass of the object, the greater the force needed to achieve the same change in motion. For any given object, a larger force causes a larger change in motion (MS-PS2-2). All positions of objects and the directions of forces and motions must be described in an arbitrarily chosen reference frame and arbitrarily chosen units of size. In order to share information with other people, these choices must also be shared (MSPS2-2). **PS1.A: Structure and properties of matter** Each atom has a charged substructure consisting of a nucleus, which is made of protons and neutrons, surrounded by electrons (HS-PS1-1) The periodic table orders elements horizontally by the number of protons in the atom's nucleus and places those with similar chemical properties in columns. The repeating patterns of this table reflect patterns of outer electron states (HS-PS1–1, HS-PS1–2) The structure and interactions of matter at the bulk scale are determined by electrical forces within and between atoms (HS-PS1–3) A stable molecule has less energy than the same set of atoms separated; one must provide at least this energy in order to take the molecule apart (HS-PS1–4)	**Cause and effect** Cause and effect relationships may be used to predict phenomena in natural or designed systems (MS-PS2-3, MS-PS2-5). **Systems and system models** Models can be used to represent systems and their interactions—such as inputs, processes, and outputs—and energy and matter flows within systems (MS-PS2-1, MS-PS2-4). **Stability and change** Explanations of stability and change in natural or designed systems can be constructed by examining the changes over time and forces at different scales (MS-PS2-2). **Patterns** Different patterns may be observed at each of the scales at which a system is studied and can provide evidence for causality in explanations of phenomena (HS-PS1-1, HS-PS1-2, HS-PS1-3, HS-PS1-5) ***Connections to engineering, technology, and applications of science*** Influence of science, **engineering, and technology on society and the natural world** The uses of technologies and any limitations on their use are driven by individual or societal needs, desires, and values; by the findings of scientific research; and by differences in such factors as climate, natural resources, and economic conditions (MS-PS2-1)

Science and engineering practices	Disciplinary core ideas	Crosscutting concepts
MS-PS2 Motion and Stability: Forces and Interactions *The performance expectations marked with an asterisk integrate traditional science content with engineering through a Practice or Disciplinary Core Idea. The section entitled "Disciplinary Core Ideas" is reproduced verbatim from A Framework for K-12 Science Education: Practices, Cross-Cutting Concepts, and Core Ideas. Integrated and reprinted with permission from the National Academy of Sciences **RST.6-8.1** Cite specific textual evidence to support analysis of science and technical texts, attending to the precise details of explanations or descriptions (MS-PS2-1, MSPS2-3) **RST.6-8.3** Follow precisely a multistep procedure when carrying out experiments, taking measurements, or performing technical tasks (MS-PS2-1, MS-PS2-2, MS-PS2-5) **WHST.6-8.1** Write arguments focused on discipline-specific content (MS-PS2-4) **WHST.6-8.7** Conduct short research projects to answer a question (including a self-generated question), drawing on several sources and generating additional related, focused questions that allow for multiple avenues of exploration (MS-PS2-1, MS-PS2-2, MS-PS2-5) **Mathematics:** **MP.2** Reason abstractly and quantitatively (MS-PS2-1, MS-PS2-2, MS-PS2-3) **6.NS.C.5** Understand that positive and negative numbers are used together to describe quantities having opposite directions or values; use positive and negative numbers to represent quantities in real-world contexts, explaining the meaning of 0 in each situation (MS-PS2-1) **6.EE.A.2** Write, read, and evaluate expressions in which letters stand for numbers (MS-PS2-1, MS-PS2-2) **7.EE.B.3** Solve multi-step real-life and mathematical problems posed with positive and negative rational numbers in any form, using tools strategically. Apply properties of operations to calculate with numbers in any form; convert between forms as appropriate; and assess the reasonableness of answers using mental computation and estimation strategies (MS-PS2-1, MS-PS2-2) **7.EE.B.4** Use variables to represent quantities in a real-world or mathematical problem, and construct simple equations and inequalities to solve problems by reasoning about the quantities (MS-PS2-1, MS-PS2-2)		

[93]Common Core Standards

CCSS.ELA-LITERACY.RST.11-12.4
Determine the meaning of symbols, key terms, and other domain-specific words and phrases as they are used in a specific scientific or technical context relevant to *grades 11–12 texts and topics.*

CCSS.ELA-LITERACY.RST.11-12.9
Synthesize information from a range of sources (e.g., texts, experiments, simulations) into a coherent understanding of a process, phenomenon, or concept, resolving conflicting information when possible.

CCSS.ELA-LITERACY.RST.11-12.10
By the end of grade 12, read and comprehend science/technical texts in the grades 11-CCR text complexity band independently and proficiently.

CCSS.ELA-LITERACY.RST.6-8.4
Determine the meaning of symbols, key terms, and other domain-specific words and phrases as they are used in a specific scientific or technical context relevant to grades 6–8 texts and topics.

Chapter 12

[92]Standards

Standards met in the chapter based on **NGSS (New Generation Science Standards)** are:

MS-PS1-4. Develop a model that predicts and describes changes in particle motion, temperature, and state of a pure substance when thermal energy is added or removed. [Clarification Statement: Emphasis is on qualitative molecular-level models of solids, liquids, and gases to show that adding or removing thermal energy increases or decreases kinetic energy of the particles until a change of state occurs. Examples of models could include drawings and diagrams. Examples of particles could include molecules or inert atoms. Examples of pure substances could include water, carbon dioxide, and helium.]

MS-PS2-3. Ask questions about data to determine the factors that affect the strength of electric and magnetic forces. [Clarification Statement: Examples of devices that use electric and magnetic forces could include electromagnets, electric motors, or generators. Examples of data could include the effect of the number of turns of wire on the strength of an electromagnet, or the effect of increasing the number or strength of magnets on the speed of an electric motor.] [Assessment Boundary: Assessment about questions that require quantitative answers is limited to proportional reasoning and algebraic thinking.]

MS-PS3-3. Apply scientific principles to design, construct, and test a device that either minimizes or maximizes thermal energy transfer.* [Clarification Statement: Examples of devices could include an insulated box, a solar cooker, and a Styrofoam cup.] [Assessment Boundary: Assessment does not include calculating the total amount of thermal energy transferred.]

HS-PS1-5. Apply scientific principles and evidence to provide an explanation about the effects of changing the temperature or concentration of the reacting particles on the rate at which a reaction occurs. [Clarification Statement: Emphasis is on student reasoning that focuses on the number and energy of collisions between molecules.] [Assessment Boundary: Assessment is limited to simple reactions in which there are only two reactants; evidence from temperature, concentration, and rate data; and qualitative relationships between rate and temperature.]

HS-PS1-3. Plan and conduct an investigation to gather evidence to compare the structure of substances at the bulk scale to infer the strength of electrical forces between particles. [Clarification Statement: Emphasis is on understanding the strengths of forces between particles, not on naming specific intermolecular forces (such as dipole-dipole). Examples of particles could include ions, atoms, molecules, and networked materials (such as graphite). Examples of bulk properties of substances could include the melting point and boiling point, vapor pressure, and surface tension.] [Assessment Boundary: Assessment does not include Raoult's law calculations of vapor pressure.]

HS-PS1-5. Apply scientific principles and evidence to provide an explanation about the effects of changing the temperature or concentration of the reacting particles on the rate at which a reaction occurs. [Clarification Statement: Emphasis is on student reasoning that focuses on the number and energy of collisions between molecules.] [Assessment Boundary: Assessment is limited to simple reactions in which there are only two reactants; evidence from temperature, concentration, and rate data; and qualitative relationships between rate and temperature.]

Science and engineering practices	Disciplinary core ideas	Crosscutting concepts
Asking questions and defining problems Asking questions and defining problems in grades 6–8 builds from grades K–5 experiences and progresses to specifying relationships between variables, and clarifying arguments and models Ask questions that can be investigated within the scope of the classroom, outdoor environment, and museums and other public facilities with available resources and, when appropriate, frame a hypothesis based on observations and scientific principles. (MS-PS2-3). **Planning and carrying out investigations** Planning and carrying out investigations to answer questions or test solutions to problems in 6–8 builds on K–5 experiences and progresses to include investigations that use multiple variables and provide evidence to support explanations or design solutions. Plan an investigation individually and collaboratively, and in the design: Identify independent and dependent variables and controls, what tools are needed to do the gathering, how measurements will be recorded, and how many data are needed to support a claim. (MS-PS2-2). **Planning and carrying out investigations** Planning and carrying out investigations in 9–12 builds on K-8 experiences and progresses to include investigations that provide evidence for and test conceptual, mathematical, physical, and empirical models. Plan and conduct an investigation individually and collaboratively to produce data to serve as the basis for evidence, and in the design: Decide on types, how much, and accuracy of data needed to produce reliable measurements and consider limitations on the precision of the data (e.g., number of trials, cost, risk, time), and refine the design accordingly. (HS-PS1-3) **Constructing explanations and designing solutions** Constructing explanations and designing solutions in 6–8 builds on K–5 experiences and progresses to include constructing explanations and designing solutions supported by multiple sources of evidence consistent with scientific ideas, principles, and theories. Apply scientific ideas or principles to design an object, tool, process or system. (MS-PS2-1). **Developing and using models** Modeling in 9–12 builds on K–8 and progresses to using, synthesizing, and developing models to predict and show relationships among variables between systems and their components in the natural and designed worlds. Use a model to predict the relationships between systems or between components of a system. (HS-PS1-1)	**PS2.A: Forces and motion** For any pair of interacting objects, the force exerted by the first object on the second object is equal in strength to the force that the second object exerts on the first, but in the opposite direction (Newton's Third Law) (MS-PS2-1). The motion of an object is determined by the sum of the forces acting on it; if the total force on the object is not zero, its motion will change. The greater the mass of the object, the greater the force needed to achieve the same change in motion. For any given object, a larger force causes a larger change in motion (MS-PS2-2). All positions of objects and the directions of forces and motions must be described in an arbitrarily chosen reference frame and arbitrarily chosen units of size. In order to share information with other people, these choices must also be shared (MSPS2-2). **PS1.A: Structure and properties of matter** Each atom has a charged substructure consisting of a nucleus, which is made of protons and neutrons, surrounded by electrons (HS-PS1-1) The periodic table orders elements horizontally by the number of protons in the atom's nucleus and places those with similar chemical properties in columns. The repeating patterns of this table reflect patterns of outer electron states (HS-PS1–1, HS-PS1–2) The structure and interactions of matter at the bulk Scale are determined by electrical forces within and between atoms. (HS-PS1–3) A stable molecule has less energy than the same set of atoms separated; one must provide at least this energy in order to take the molecule apart (HS-PS1–4)	**Cause and effect** Cause and effect relationships may be used to predict phenomena in natural or designed systems (MS-PS2-3, MS-PS2-5). **Systems and system models** Models can be used to represent systems and their interactions—such as inputs, processes, and outputs—and energy and matter flows within systems (MS-PS2-1, MS-PS2-4). **Stability and change** Explanations of stability and change in natural or designed systems can be constructed by examining the changes over time and forces at different scales (MS-PS2-2). **Patterns** Different patterns may be observed at each of the scales at which a system is studied and can provide evidence for causality in explanations of phenomena (HS-PS1-1, HS-PS1-2, HS-PS1-3, HS-PS1-5) ***Connections to engineering, technology, and applications of science*** **Influence of science, engineering, and technology on society and the natural world** The uses of technologies and any limitations on their use are driven by individual or societal needs, desires, and values; by the findings of scientific research; and by differences in such factors as climate, natural resources, and economic conditions (MS-PS2-1)

Science and engineering practices	Disciplinary core ideas	Crosscutting concepts
MS-PS2 Motion and Stability: Forces and Interactions *The performance expectations marked with an asterisk integrate traditional science content with engineering through a Practice or Disciplinary Core Idea. The section entitled "Disciplinary Core Ideas" is reproduced verbatim from A Framework for K-12 Science Education: Practices, Cross-Cutting Concepts, and Core Ideas. Integrated and reprinted with permission from the National Academy of Sciences. **RST.6-8.1** Cite specific textual evidence to support analysis of science and technical texts, attending to the precise details of explanations or descriptions (MS-PS2-1, MSPS2-3) **RST.6-8.3** Follow precisely a multistep procedure when carrying out experiments, taking measurements, or performing technical tasks (MS-PS2-1, MS-PS2-2, MS-PS2-5) **WHST.6-8.1** Write arguments focused on discipline-specific content (MS-PS2-4) **WHST.6-8.7** Conduct short research projects to answer a question (including a self-generated question), drawing on several sources and generating additional related, focused questions that allow for multiple avenues of exploration (MS-PS2-1, MS-PS2-2, MS-PS2-5) **Mathematics:** **MP.2** Reason abstractly and quantitatively (MS-PS2-1, MS-PS2-2, MS-PS2-3) **6.NS.C.5** Understand that positive and negative numbers are used together to describe quantities having opposite directions or values; use positive and negative numbers to represent quantities in real-world contexts, explaining the meaning of 0 in each situation (MS-PS2-1) **6.EE.A.2** Write, read, and evaluate expressions in which letters stand for numbers (MS-PS2-1, MS-PS2-2) **7.EE.B.3** Solve multi-step real-life and mathematical problems posed with positive and negative rational numbers in any form, using tools strategically. Apply properties of operations to calculate with numbers in any form; convert between forms as appropriate; and assess the reasonableness of answers using mental computation and estimation strategies (MS-PS2-1, MS-PS2-2) **7.EE.B.4** Use variables to represent quantities in a real-world or mathematical problem, and construct simple equations and inequalities to solve problems by reasoning about the quantities (MS-PS2-1, MS-PS2-2)		

[93]Common Core Standards

CCSS.ELA-LITERACY.RST.11-12.4
Determine the meaning of symbols, key terms, and other domain-specific words and phrases as they are used in a specific scientific or technical context relevant to *grades 11–12 texts and topics.*

CCSS.ELA-LITERACY.RST.11-12.9
Synthesize information from a range of sources (e.g., texts, experiments, simulations) into a coherent understanding of a process, phenomenon, or concept, resolving conflicting information when possible.

CCSS.ELA-LITERACY.RST.11-12.10
By the end of grade 12, read and comprehend science/technical texts in the grades 11-CCR text complexity band independently and proficiently.

CCSS.ELA-LITERACY.RST.6-8.4
Determine the meaning of symbols, key terms, and other domain-specific words and phrases as they are used in a specific scientific or technical context relevant to grades 6–8 texts and topics.

Chapter 13

[92]Standards

Standards met in the chapter based on **NGSS (New Generation Science Standards)** are:

MS-PS1-4. Develop a model that predicts and describes changes in particle motion, temperature, and state of a pure substance when thermal energy is added or removed. [Clarification Statement: Emphasis is on qualitative molecular-level models of solids, liquids, and gases to show that adding or removing thermal energy increases or decreases kinetic energy of the particles until a change of state occurs. Examples of models could include drawings and diagrams. Examples of particles could include molecules or inert atoms. Examples of pure substances could include water, carbon dioxide, and helium.]

MS-PS2-3. Ask questions about data to determine the factors that affect the strength of electric and magnetic forces. [Clarification Statement: Examples of devices that use electric and magnetic forces could include electromagnets, electric motors, or generators. Examples of data could include the effect of the number of turns of wire on the strength of an electromagnet, or the effect of increasing the number or strength of magnets on the speed of an electric motor.] [Assessment Boundary: Assessment about questions that require quantitative answers is limited to proportional reasoning and algebraic thinking.]

MS-PS3-3. Apply scientific principles to design, construct, and test a device that either minimizes or maximizes thermal energy transfer.* [Clarification Statement: Examples of devices could include an insulated box, a solar cooker, and a Styrofoam cup.] [Assessment Boundary: Assessment does not include calculating the total amount of thermal energy transferred.]

HS-PS1-5. Apply scientific principles and evidence to provide an explanation about the effects of changing the temperature or concentration of the reacting particles on the rate at which a reaction occurs. [Clarification Statement: Emphasis is on student reasoning that focuses on the number and energy of collisions between molecules.] [Assessment Boundary: Assessment is limited to simple reactions in which there are only two reactants; evidence from temperature, concentration, and rate data; and qualitative relationships between rate and temperature.]

HS-PS1-3. Plan and conduct an investigation to gather evidence to compare the structure of substances at the bulk scale to infer the strength of electrical forces between particles. [Clarification Statement: Emphasis is on understanding the strengths of forces between particles, not on naming specific intermolecular forces (such as dipole-dipole). Examples of particles could include ions, atoms, molecules, and networked materials (such as graphite). Examples of bulk properties of substances could include the melting point and boiling point, vapor pressure, and surface tension.] [Assessment Boundary: Assessment does not include Raoult's law calculations of vapor pressure.]

HS-PS1-5. Apply scientific principles and evidence to provide an explanation about the effects of changing the temperature or concentration of the reacting particles on the rate at which a reaction occurs. [Clarification Statement: Emphasis is on student reasoning that focuses on the number and energy of collisions between molecules.] [Assessment Boundary: Assessment is limited to simple reactions in which there are only two reactants; evidence from temperature, concentration, and rate data; and qualitative relationships between rate and temperature.]

Science and engineering practices	Disciplinary core ideas	Crosscutting concepts
Asking questions and defining problems Asking questions and defining problems in grades 6–8 builds from grades K–5 experiences and progresses to specifying relationships between variables, and clarifying arguments and models Ask questions that can be investigated within the scope of the classroom, outdoor environment, and museums and other public facilities with available resources and, when appropriate, frame a hypothesis based on observations and scientific principles (MS-PS2-3). **Planning and carrying out investigations** Planning and carrying out investigations to answer questions or test solutions to problems in 6–8 builds on K–5 experiences and progresses to include investigations that use multiple variables and provide evidence to support explanations or design solutions. Plan an investigation individually and collaboratively, and in the design: Identify independent and dependent variables and controls, what tools are needed to do the gathering, how measurements will be recorded, and how many data are needed to support a claim (MS-PS2-2). **Planning and carrying out investigations** Planning and carrying out investigations in 9–12 builds on K-8 experiences and progresses to include investigations that provide evidence for and test conceptual, mathematical, physical, and empirical models Plan and conduct an investigation individually and collaboratively to produce data to serve as the basis for evidence, and in the design: Decide on types, how much, and accuracy of data needed to produce reliable measurements and consider limitations on the precision of the data (e.g., number of trials, cost, risk, time), and refine the design accordingly (HS-PS1-3) **Constructing explanations and designing solutions** Constructing explanations and designing solutions in 6–8 builds on K–5 experiences and progresses to include constructing explanations and designing solutions supported by multiple sources of evidence consistent with scientific ideas, principles, and theories Apply scientific ideas or principles to design an object, tool, process or system (MS-PS2-1). **Developing and using models** Modeling in 9–12 builds on K–8 and progresses to using, synthesizing, and developing models to predict and show relationships among variables between systems and their components in the natural and designed worlds Use a model to predict the relationships between systems or between components of a system (HS-PS1-1)	**PS2.A: Forces and motion** For any pair of interacting objects, the force exerted by the first object on the second object is equal in strength to the force that the second object exerts on the first, but in the opposite direction (Newton's Third Law) (MS-PS2-1). The motion of an object is determined by the sum of the forces acting on it; if the total force on the object is not zero, its motion will change. The greater the mass of the object, the greater the force needed to achieve the same change in motion. For any given object, a larger force causes a larger change in motion (MS-PS2-2). All positions of objects and the directions of forces and motions must be described in an arbitrarily chosen reference frame and arbitrarily chosen units of size. In order to share information with other people, these choices must also be shared (MSPS2-2). **PS1.A: Structure and properties of matter** Each atom has a charged substructure consisting of a nucleus, which is made of protons and neutrons, surrounded by electrons (HS-PS1-1) The periodic table orders elements horizontally by the number of protons in the atom's nucleus and places those with similar chemical properties in columns. The repeating patterns of this table reflect patterns of outer electron states (HS-PS1-1, HS-PS1-2) The structure and interactions of matter at the bulk scale are determined by electrical forces within and between atoms. (HS-PS1-3) A stable molecule has less energy than the same set of atoms separated; one must provide at least this energy in order to take the molecule apart (HS-PS1-4)	**Cause and effect** Cause and effect relationships may be used to predict phenomena in natural or designed systems (MS-PS2-3, MS-PS2-5). **Systems and system models** Models can be used to represent systems and their interactions—such as inputs, processes, and outputs—and energy and matter flows within systems (MS-PS2-1, MS-PS2-4). **Stability and change** Explanations of stability and change in natural or designed systems can be constructed by examining the changes over time and forces at different scales (MS-PS2-2). **Patterns** Different patterns may be observed at each of the scales at which a system is studied and can provide evidence for causality in explanations of phenomena. (HS-PS1-1, HS-PS1-2, HS-PS1-3, HS-PS1-5) ***Connections to engineering, technology, and applications of science*** **Influence of science, engineering, and technology on society and the natural world** The uses of technologies and any limitations on their use are driven by individual or societal needs, desires, and values; by the findings of scientific research; and by differences in such factors as climate, natural resources, and economic conditions (MS-PS2-1)

Science and engineering practices	Disciplinary core ideas	Crosscutting concepts
MS-PS2 Motion and Stability: Forces and Interactions *The performance expectations marked with an asterisk integrate traditional science content with engineering through a Practice or Disciplinary Core Idea. The section entitled "Disciplinary Core Ideas" is reproduced verbatim from A Framework for K-12 Science Education: Practices, Cross-Cutting Concepts, and Core Ideas. Integrated and reprinted with permission from the National Academy of Sciences. **RST.6-8.1** Cite specific textual evidence to support analysis of science and technical texts, attending to the precise details of explanations or descriptions (MS-PS2-1, MSPS2-3) **RST.6-8.3** Follow precisely a multistep procedure when carrying out experiments, taking measurements, or performing technical tasks (MS-PS2-1, MS-PS2-2, MS-PS2-5) **WHST.6-8.1** Write arguments focused on discipline-specific content (MS-PS2-4) **WHST.6-8.7** Conduct short research projects to answer a question (including a self-generated question), drawing on several sources and generating additional related, focused questions that allow for multiple avenues of exploration (MS-PS2-1, MS-PS2-2, MS-PS2-5) **Mathematics:** **MP.2** Reason abstractly and quantitatively (MS-PS2-1, MS-PS2-2, MS-PS2-3) **6.NS.C.5** Understand that positive and negative numbers are used together to describe quantities having opposite directions or values; use positive and negative numbers to represent quantities in real-world contexts, explaining the meaning of 0 in each situation (MS-PS2-1) **6.EE.A.2** Write, read, and evaluate expressions in which letters stand for numbers (MS-PS2-1, MS-PS2-2) **7.EE.B.3** Solve multi-step real-life and mathematical problems posed with positive and negative rational numbers in any form, using tools strategically. Apply properties of operations to calculate with numbers in any form; convert between forms as appropriate; and assess the reasonableness of answers using mental computation and estimation strategies (MS-PS2-1, MS-PS2-2) **7.EE.B.4** Use variables to represent quantities in a real-world or mathematical problem, and construct simple equations and inequalities to solve problems by reasoning about the quantities (MS-PS2-1, MS-PS2-2)		

[93]Common Core Standards

CCSS.ELA-LITERACY.RST.11-12.4
Determine the meaning of symbols, key terms, and other domain-specific words and phrases as they are used in a specific scientific or technical context relevant to *grades 11–12 texts and topics.*

CCSS.ELA-LITERACY.RST.11-12.9
Synthesize information from a range of sources (e.g., texts, experiments, simulations) into a coherent understanding of a process, phenomenon, or concept, resolving conflicting information when possible.

CCSS.ELA-LITERACY.RST.11-12.10
By the end of grade 12, read and comprehend science/technical texts in the grades 11-CCR text complexity band independently and proficiently.

CCSS.ELA-LITERACY.RST.6-8.4
Determine the meaning of symbols, key terms, and other domain-specific words and phrases as they are used in a specific scientific or technical context relevant to grades 6–8 texts and topics.

Chapter 14

[92]Standards

Standards met in the chapter based on **NGSS (New Generation Science Standards)** are:

MS-PS1-4. Develop a model that predicts and describes changes in particle motion, temperature, and state of a pure substance when thermal energy is added or removed. [Clarification Statement: Emphasis is on qualitative molecular-level models of solids, liquids, and gases to show that adding or removing thermal energy increases or decreases kinetic energy of the particles until a change of state occurs. Examples of models could include drawings and diagrams. Examples of particles could include molecules or inert atoms. Examples of pure substances could include water, carbon dioxide, and helium.]

MS-PS3-3. Apply scientific principles to design, construct, and test a device that either minimizes or maximizes thermal energy transfer.* [Clarification Statement: Examples of devices could include an insulated box, a solar cooker, and a Styrofoam cup.] [Assessment Boundary: Assessment does not include calculating the total amount of thermal energy transferred.]

HS-PS1-5. Apply scientific principles and evidence to provide an explanation about the effects of changing the temperature or concentration of the reacting particles on the rate at which a reaction occurs. [Clarification Statement: Emphasis is on student reasoning that focuses on the number and energy of collisions between molecules.] [Assessment Boundary: Assessment is limited to simple reactions in which there are only two reactants; evidence from temperature, concentration, and rate data; and qualitative relationships between rate and temperature.]

Science and engineering practices	Disciplinary core ideas	Crosscutting concepts
Asking questions and defining problems Asking questions and defining problems in grades 6–8 builds from grades K–5 experiences and progresses to specifying relationships between variables, and clarifying arguments and models Ask questions that can be investigated within the scope of the classroom, outdoor environment, and museums and other public facilities with available resources and, when appropriate, frame a hypothesis based on observations and scientific principles (MS-PS2-3). **Planning and carrying out investigations** Planning and carrying out investigations in 9–12 builds on K-8 experiences and progresses to include investigations that provide evidence for and test conceptual, mathematical, physical, and empirical models Plan and conduct an investigation individually and collaboratively to produce data to serve as the basis for evidence, and in the design: decide on types, how much, and accuracy of data needed to produce reliable measurements and consider limitations on the precision of the data (e.g., number of trials, cost, risk, time), and refine the design accordingly (HS-PS1-3) **Constructing explanations and designing solutions** Constructing explanations and designing solutions in 6–8 builds on K–5 experiences and progresses to include constructing explanations and designing solutions supported by multiple sources of evidence consistent with scientific ideas, principles, and theories. Apply scientific ideas or principles to design an object, tool, process or system (MS-PS2-1). **Developing and using models** Modeling in 9–12 builds on K–8 and progresses to using, synthesizing, and developing models to predict and show relationships among variables between systems and their components in the natural and designed worlds Use a model to predict the relationships between systems or between components of a system (HS-PS1-1)	**PS2.A: Forces and motion** The motion of an object is determined by the sum of the forces acting on it; if the total force on the object is not zero, its motion will change. The greater the mass of the object, the greater the force needed to achieve the same change in motion. For any given object, a larger force causes a larger change in motion (MS-PS2-2). All positions of objects and the directions of forces and motions must be described in an arbitrarily chosen reference frame and arbitrarily chosen units of size. In order to share information with other people, these choices must also be shared (MSPS2-2). **PS1.A: Structure and properties of matter** The periodic table orders elements horizontally by the number of protons in the atom's nucleus and places those with similar chemical properties in columns. The repeating patterns of this table reflect patterns of outer electron states (HS-PS1-1, HS-PS1-2) The structure and interactions of matter at the bulk scale are determined by electrical forces within and between atoms (HS-PS1-3) A stable molecule has less energy than the same set of atoms separated; one must provide at least this energy in order to take the molecule apart (HS-PS1-4)	**Cause and effect** Cause and effect relationships may be used to predict phenomena in natural or designed systems (MS-PS2-3, MS-PS2-5). **Systems and system models** Models can be used to represent systems and their interactions—such as inputs, processes, and outputs—and energy and matter flows within systems (MS-PS2-1, MS-PS2-4). **Stability and change** Explanations of stability and change in natural or designed systems can be constructed by examining the changes over time and forces at different scales (MS-PS2-2). **Patterns** Different patterns may be observed at each of the scales at which a system is studied and can provide evidence for causality in explanations of phenomena (HS-PS1-1, HS-PS1-2, HS-PS1-3, HS-PS1-5) ***Connections to engineering, technology, and applications of science*** **Influence of science, engineering, and technology on society and the natural world** The uses of technologies and any limitations on their use are driven by individual or societal needs, desires, and values; by the findings of scientific research; and by differences in such factors as climate, natural resources, and economic conditions (MS-PS2-1)

Science and engineering practices	Disciplinary core ideas	Crosscutting concepts
MS-PS2 Motion and Stability: Forces and Interactions *The performance expectations marked with an asterisk integrate traditional science content with engineering through a Practice or Disciplinary Core Idea. The section entitled "Disciplinary Core Ideas" is reproduced verbatim from A Framework for K-12 Science Education: Practices, Cross-Cutting Concepts, and Core Ideas. Integrated and reprinted with permission from the National Academy of Sciences. **RST.6-8.1** Cite specific textual evidence to support analysis of science and technical texts, attending to the precise details of explanations or descriptions (MS-PS2-1, MSPS2-3) **RST.6-8.3** Follow precisely a multistep procedure when carrying out experiments, taking measurements, or performing technical tasks (MS-PS2-1, MS-PS2-2, MS-PS2-5) **WHST.6-8.1** Write arguments focused on discipline-specific content (MS-PS2-4) **WHST.6-8.7** Conduct short research projects to answer a question (including a self-generated question), drawing on several sources and generating additional related, focused questions that allow for multiple avenues of exploration (MS-PS2-1, MS-PS2-2, MS-PS2-5) **Mathematics:** **MP.2** Reason abstractly and quantitatively (MS-PS2-1, MS-PS2-2, MS-PS2-3) **6.NS.C.5** Understand that positive and negative numbers are used together to describe quantities having opposite directions or values; use positive and negative numbers to represent quantities in real-world contexts, explaining the meaning of 0 in each situation (MS-PS2-1) **6.EE.A.2** Write, read, and evaluate expressions in which letters stand for numbers (MS-PS2-1, MS-PS2-2) **7.EE.B.3** Solve multi-step real-life and mathematical problems posed with positive and negative rational numbers in any form, using tools strategically. Apply properties of operations to calculate with numbers in any form; convert between forms as appropriate; and assess the reasonableness of answers using mental computation and estimation strategies (MS-PS2-1, MS-PS2-2) **7.EE.B.4** Use variables to represent quantities in a real-world or mathematical problem, and construct simple equations and inequalities to solve problems by reasoning about the quantities (MS-PS2-1, MS-PS2-2)		

[93]Common Core Standards

CCSS.ELA-LITERACY.RST.11-12.4
Determine the meaning of symbols, key terms, and other domain-specific words and phrases as they are used in a specific scientific or technical context relevant to *grades 11–12 texts and topics.*

CCSS.ELA-LITERACY.RST.11-12.9
Synthesize information from a range of sources (e.g., texts, experiments, simulations) into a coherent understanding of a process, phenomenon, or concept, resolving conflicting information when possible.

CCSS.ELA-LITERACY.RST.11-12.10
By the end of grade 12, read and comprehend science/technical texts in the grades 11-CCR text complexity band independently and proficiently.

CCSS.ELA-LITERACY.RST.6-8.4
Determine the meaning of symbols, key terms, and other domain-specific words and phrases as they are used in a specific scientific or technical context relevant to grades 6–8 texts and topics.

Chapter 15

[92]Standards

Standards met in the chapter based on **NGSS (New Generation Science Standards)** are:

MS-PS3-3. Apply scientific principles to design, construct, and test a device that either minimizes or maximizes thermal energy transfer.* [Clarification Statement: Examples of devices could include an insulated box, a solar cooker, and a Styrofoam cup.] [Assessment Boundary: Assessment does not include calculating the total amount of thermal energy transferred.]

MS-PS2-2. Plan an investigation to provide evidence that the change in an object's motion depends on the sum of the forces on the object and the mass of the object. [Clarification Statement: Emphasis is on balanced (Newton's First Law) and unbalanced forces in a system, qualitative comparisons of forces, mass and changes in motion (Newton's Second Law), frame of reference, and specification of units.] [Assessment Boundary: Assessment is limited to forces and changes in motion in one-dimension in an inertial reference frame and to change in one variable at a time. Assessment does not include the use of trigonometry.]

HS-PS2-2. Use mathematical representations to support the claim that the total momentum of a system of objects is conserved when there is no net force on the system. [Clarification Statement: Emphasis is on the quantitative conservation of momentum in interactions and the qualitative meaning of this principle.] [Assessment Boundary: Assessment is limited to systems of two macroscopic bodies moving in one dimension.]

HS-PS2-3. Apply scientific and engineering ideas to design, evaluate, and refine a device that minimizes the force on a macroscopic object during a collision.* [Clarification Statement: Examples of evaluation and refinement could include determining the success of the device at protecting an object from damage and modifying the design to improve it. Examples of a device could include a football helmet or a parachute.] [Assessment Boundary: Assessment is limited to qualitative evaluations and/or algebraic manipulations.]

Science and engineering practices	Disciplinary core ideas	Crosscutting concepts
Asking questions and defining problems Asking questions and defining problems in grades 6–8 builds from grades K–5 experiences and progresses to specifying relationships between variables, and clarifying arguments and models Ask questions that can be investigated within the scope of the classroom, outdoor environment, and museums and other public facilities with available resources and, when appropriate, frame a hypothesis based on observations and scientific principles (MS-PS2-3). **Planning and carrying out investigations** Planning and carrying out investigations in 9–12 builds on K-8 experiences and progresses to include investigations that provide evidence for and test conceptual, mathematical, physical, and empirical models Plan and conduct an investigation individually and collaboratively to produce data to serve as the basis for evidence, and in the design: Decide on types, how much, and accuracy of data needed to produce reliable measurements and consider limitations on the precision of the data (e.g., number of trials, cost, risk, time), and refine the design accordingly (HS-PS1-3) **Constructing explanations and designing solutions** Constructing explanations and designing solutions in 6–8 builds on K–5 experiences and progresses to include constructing explanations and designing solutions supported by multiple sources of evidence consistent with scientific ideas, principles, and theories. Apply scientific ideas or principles to design an object, tool, process or system (MS-PS2-1). **Developing and using models** Modeling in 9–12 builds on K–8 and progresses to using, synthesizing, and developing models to predict and show relationships among variables between systems and their components in the natural and designed worlds Use a model to predict the relationships between systems or between components of a system (HS-PS1-1)	**PS2.A: Forces and motion** The motion of an object is determined by the sum of the forces acting on it; if the total force on the object is not zero, its motion will change. The greater the mass of the object, the greater the force needed to achieve the same change in motion. For any given object, a larger force causes a larger change in motion (MS-PS2-2). All positions of objects and the directions of forces and motions must be described in an arbitrarily chosen reference frame and arbitrarily chosen units of size. In order to share information with other people, these choices must also be shared (MSPS2-2). **PS1.A: Structure and properties of matter** The periodic table orders elements horizontally by the number of protons in the atom's nucleus and places those with similar chemical properties in columns. The repeating patterns of this table reflect patterns of outer electron states (HS-PS1-1, HS-PS1-2) The structure and interactions of matter at the bulk scale are determined by electrical forces within and between atoms (HS-PS1-3) A stable molecule has less energy than the same set of atoms separated; one must provide at least this energy in order to take the molecule apart (HS-PS1-4)	**Cause and effect** Cause and effect relationships may be used to predict phenomena in natural or designed systems (MS-PS2-3, MS-PS2-5). **Systems and system models** Models can be used to represent systems and their interactions—such as inputs, processes, and outputs—and energy and matter flows within systems (MS-PS2-1, MS-PS2-4). **Stability and change** Explanations of stability and change in natural or designed systems can be constructed by examining the changes over time and forces at different scales (MS-PS2-2). **Patterns** Different patterns may be observed at each of the scales at which a system is studied and can provide evidence for causality in explanations of phenomena (HS-PS1-1, HS-PS1-2, HS-PS1-3, HS-PS1-5) ***Connections to engineering, technology, and applications of science*** **Influence of science, engineering, and technology on society and the natural world** The uses of technologies and any limitations on their use are driven by individual or societal needs, desires, and values; by the findings of scientific research; and by differences in such factors as climate, natural resources, and economic conditions (MS-PS2-1)

<table>
<tr><th>Science and engineering practices</th><th>Disciplinary core ideas</th><th>Crosscutting concepts</th></tr>
<tr><td colspan="3">MS-PS2 Motion and Stability: Forces and Interactions *The performance expectations marked with an asterisk integrate traditional science content with engineering through a Practice or Disciplinary Core Idea. The section entitled "Disciplinary Core Ideas" is reproduced verbatim from A Framework for K-12 Science Education: Practices, Cross-Cutting Concepts, and Core Ideas. Integrated and reprinted with permission from the National Academy of Sciences.
RST.6-8.1 Cite specific textual evidence to support analysis of science and technical texts, attending to the precise details of explanations or descriptions (MS-PS2-1, MSPS2-3)
RST.6-8.3 Follow precisely a multistep procedure when carrying out experiments, taking measurements, or performing technical tasks (MS-PS2-1, MS-PS2-2, MS-PS2-5)
WHST.6-8.1 Write arguments focused on discipline-specific content (MS-PS2-4)
WHST.6-8.7 Conduct short research projects to answer a question (including a self-generated question), drawing on several sources and generating additional related, focused questions that allow for multiple avenues of exploration (MS-PS2-1, MS-PS2-2, MS-PS2-5)
Mathematics:
MP.2 Reason abstractly and quantitatively (MS-PS2-1, MS-PS2-2, MS-PS2-3)
6.NS.C.5 Understand that positive and negative numbers are used together to describe quantities having opposite directions or values; use positive and negative numbers to represent quantities in real-world contexts, explaining the meaning of 0 in each situation (MS-PS2-1)
6.EE.A.2 Write, read, and evaluate expressions in which letters stand for numbers (MS-PS2-1, MS-PS2-2)
7.EE.B.3 Solve multi-step real-life and mathematical problems posed with positive and negative rational numbers in any form, using tools strategically. Apply properties of operations to calculate with numbers in any form; convert between forms as appropriate; and assess the reasonableness of answers using mental computation and estimation strategies (MS-PS2-1, MS-PS2-2)
7.EE.B.4 Use variables to represent quantities in a real-world or mathematical problem, and construct simple equations and inequalities to solve problems by reasoning about the quantities (MS-PS2-1, MS-PS2-2)</td></tr>
</table>

[93]Common Core Standards

CCSS.ELA-LITERACY.RST.11-12.4
Determine the meaning of symbols, key terms, and other domain-specific words and phrases as they are used in a specific scientific or technical context relevant to *grades 11–12 texts and topics.*

CCSS.ELA-LITERACY.RST.11-12.9
Synthesize information from a range of sources (e.g., texts, experiments, simulations) into a coherent understanding of a process, phenomenon, or concept, resolving conflicting information when possible.

CCSS.ELA-LITERACY.RST.11-12.10
By the end of grade 12, read and comprehend science/technical texts in the grades 11-CCR text complexity band independently and proficiently.

CCSS.ELA-LITERACY.RST.6-8.4
Determine the meaning of symbols, key terms, and other domain-specific words and phrases as they are used in a specific scientific or technical context relevant to grades 6–8 texts and topics.

Chapter 16

[92]Standards

Standards met in the chapter based on **NGSS (New Generation Science Standards)** are:

MS-PS3-3. Apply scientific principles to design, construct, and test a device that either minimizes or maximizes thermal energy transfer.* [Clarification Statement: Examples of devices could include an insulated box, a solar cooker, and a Styrofoam cup.] [Assessment Boundary: Assessment does not include calculating the total amount of thermal energy transferred.]

MS-PS2-2. Plan an investigation to provide evidence that the change in an object's motion depends on the sum of the forces on the object and the mass of the object. [Clarification Statement: Emphasis is on balanced (Newton's First Law) and unbalanced forces in a system, qualitative comparisons of forces, mass and changes in motion (Newton's Second Law), frame of reference, and specification of units.] [Assessment Boundary: Assessment is limited to forces and changes in motion in one-dimension in an inertial reference frame and to change in one variable at a time. Assessment does not include the use of trigonometry.]

HS-PS2-2. Use mathematical representations to support the claim that the total momentum of a system of objects is conserved when there is no net force on the system. [Clarification Statement: Emphasis is on the quantitative conservation of momentum in interactions and the qualitative meaning of this principle.] [Assessment Boundary: Assessment is limited to systems of two macroscopic bodies moving in one dimension.]

HS-PS2-3. Apply scientific and engineering ideas to design, evaluate, and refine a device that minimizes the force on a macroscopic object during a collision.* [Clarification Statement: Examples of evaluation and refinement could include determining the success of the device at protecting an object from damage and modifying the design to improve it. Examples of a device could include a football helmet or a parachute.] [Assessment Boundary: Assessment is limited to qualitative evaluations and/or algebraic manipulations.]

HS-PS1-5. Apply scientific principles and evidence to provide an explanation about the effects of changing the temperature or concentration of the reacting particles on the rate at which a reaction occurs. [Clarification Statement: Emphasis is on student reasoning that focuses on the number and energy of collisions between molecules.] [Assessment Boundary: Assessment is limited to simple reactions in which there are only two reactants; evidence from temperature, concentration, and rate data; and qualitative relationships between rate and temperature.]

Science and engineering practices	Disciplinary core ideas	Crosscutting concepts
Asking questions and defining problems Asking questions and defining problems in grades 6–8 builds from grades K–5 experiences and progresses to specifying relationships between variables, and clarifying arguments and models Ask questions that can be investigated within the scope of the classroom, outdoor environment, and museums and other public facilities with available resources and, when appropriate, frame a hypothesis based on observations and scientific principles (MS-PS2-3). **Planning and carrying out investigations** Planning and carrying out investigations in 9–12 builds on K-8 experiences and progresses to include investigations that provide evidence for and test conceptual, mathematical, physical, and empirical models Plan and conduct an investigation individually and collaboratively to produce data to serve as the basis for evidence, and in the design: decide on types, how much, and accuracy of data needed to produce reliable measurements and consider limitations on the precision of the data (e.g., number of trials, cost, risk, time), and refine the design accordingly (HS-PS1-3) **Constructing explanations and designing solutions** Constructing explanations and designing solutions in 6–8 builds on K–5 experiences and progresses to include constructing explanations and designing solutions supported by multiple sources of evidence consistent with scientific ideas, principles, and theories. Apply scientific ideas or principles to design an object, tool, process or system (MS-PS2-1). **Developing and using models** Modeling in 9–12 builds on K–8 and progresses to using, synthesizing, and developing models to predict and show relationships among variables between systems and their components in the natural and designed worlds Use a model to predict the relationships between systems or between components of a system (HS-PS1-1)	**PS2.A: Forces and motion** The motion of an object is determined by the sum of the forces acting on it; if the total force on the object is not zero, its motion will change. The greater the mass of the object, the greater the force needed to achieve the same change in motion. For any given object, a larger force causes a larger change in motion (MS-PS2-2). All positions of objects and the directions of forces and motions must be described in an arbitrarily chosen reference frame and arbitrarily chosen units of size. In order to share information with other people, these choices must also be shared (MSPS2-2). **PS1.A: Structure and properties of matter** The periodic table orders elements horizontally by the number of protons in the atom's nucleus and places those with similar chemical properties in columns. The repeating patterns of this table reflect patterns of outer electron states (HS-PS1-1, HS-PS1-2) The structure and interactions of matter at the bulk scale are determined by electrical forces within and between atoms (HS-PS1-3) A stable molecule has less energy than the same set of atoms separated; one must provide at least this energy in order to take the molecule apart (HS-PS1-4)	**Cause and effect** Cause and effect relationships may be used to predict phenomena in natural or designed systems (MS-PS2-3, MS-PS2-5). **Systems and system models** Models can be used to represent systems and their interactions—such as inputs, processes, and outputs—and energy and matter flows within systems (MS-PS2-1, MS-PS2-4). **Stability and change** Explanations of stability and change in natural or designed systems can be constructed by examining the changes over time and forces at different scales (MS-PS2-2). **Patterns** Different patterns may be observed at each of the scales at which a system is studied and can provide evidence for causality in explanations of phenomena (HS-PS1-1, HS-PS1-2, HS-PS1-3, HS-PS1-5) ***Connections to engineering, technology, and applications of science*** **Influence of science, engineering, and technology on society and the natural world** The uses of technologies and any limitations on their use are driven by individual or societal needs, desires, and values; by the findings of scientific research; and by differences in such factors as climate, natural resources, and economic conditions (MS-PS2-1)

Science and engineering practices	Disciplinary core ideas	Crosscutting concepts
MS-PS2 Motion and Stability: Forces and Interactions *The performance expectations marked with an asterisk integrate traditional science content with engineering through a Practice or Disciplinary Core Idea. The section entitled "Disciplinary Core Ideas" is reproduced verbatim from A Framework for K-12 Science Education: Practices, Cross-Cutting Concepts, and Core Ideas. Integrated and reprinted with permission from the National Academy of Sciences. **RST.6-8.1** Cite specific textual evidence to support analysis of science and technical texts, attending to the precise details of explanations or descriptions (MS-PS2-1, MSPS2-3) **RST.6-8.3** Follow precisely a multistep procedure when carrying out experiments, taking measurements, or performing technical tasks (MS-PS2-1, MS-PS2-2, MS-PS2-5) **WHST.6-8.1** Write arguments focused on discipline-specific content (MS-PS2-4) **WHST.6-8.7** Conduct short research projects to answer a question (including a self-generated question), drawing on several sources and generating additional related, focused questions that allow for multiple avenues of exploration (MS-PS2-1, MS-PS2-2, MS-PS2-5) **Mathematics:** **MP.2** Reason abstractly and quantitatively (MS-PS2-1, MS-PS2-2, MS-PS2-3) **6.NS.C.5** Understand that positive and negative numbers are used together to describe quantities having opposite directions or values; use positive and negative numbers to represent quantities in real-world contexts, explaining the meaning of 0 in each situation (MS-PS2-1) **6.EE.A.2** Write, read, and evaluate expressions in which letters stand for numbers (MS-PS2-1, MS-PS2-2) **7.EE.B.3** Solve multi-step real-life and mathematical problems posed with positive and negative rational numbers in any form, using tools strategically. Apply properties of operations to calculate with numbers in any form; convert between forms as appropriate; and assess the reasonableness of answers using mental computation and estimation strategies (MS-PS2-1, MS-PS2-2) **7.EE.B.4** Use variables to represent quantities in a real-world or mathematical problem, and construct simple equations and inequalities to solve problems by reasoning about the quantities (MS-PS2-1, MS-PS2-2)		

[93]Common Core Standards

CCSS.ELA-LITERACY.RST.11-12.4
Determine the meaning of symbols, key terms, and other domain-specific words and phrases as they are used in a specific scientific or technical context relevant to *grades 11–12 texts and topics.*

CCSS.ELA-LITERACY.RST.11-12.9
Synthesize information from a range of sources (e.g., texts, experiments, simulations) into a coherent understanding of a process, phenomenon, or concept, resolving conflicting information when possible.

CCSS.ELA-LITERACY.RST.11-12.10
By the end of grade 12, read and comprehend science/technical texts in the grades 11-CCR text complexity band independently and proficiently.

CCSS.ELA-LITERACY.RST.6-8.4
Determine the meaning of symbols, key terms, and other domain-specific words and phrases as they are used in a specific scientific or technical context relevant to grades 6–8 texts and topics.

Bibliography

Amino acid. (n.d.). Retrieved March 6, 2015, from http://en.wikipedia.org/wiki/Amino_acid

APME. (n.d.). Report on Polymer Identification Codes for Packaging, www.Plasticseurope.org, Retrieved from Ref: apme identification Report 0114

Brody, Garry S. "Silicone Breast Implant Safety and Efficacy." *eMedicine*. Eds. James Neal Long, et al. 25 Apr. 2012. Medscape. 13 May 2014 <http://emedicine.medscape.com/article/1275451-overview>.

Ceramic Science-Engineering. (n.d.). Retrieved February 12, 2015, from http://ceramics.org/knowledge-center/learn-about-ceramics

Chemistry Hub. (2014, November 14). Retrieved January 25, 2015, from http://www.boundless.com/chemistry/

Corrosion Consulting Services, Corrosion Short Courses and Corrosion Expert Witness. WebCorr has NACE certified Corrosion Specialist providing corrosion consultancy, corrosion training courses and corrosion expert witness. (n.d.). Retrieved February 4, 2015, from http://www.corrosionclinic.com/index.html

Darouiche, Rabih O. "Treatment of Infections Associated with Surgical Implants." *New England Journal of Medicine* 350 14 (2004): 1422–1429. *New England Journal of Medicine*. 25 Mar. 2009 <http://content.nejm.org/cgi/content/extract/350/14/1422>.

Fiberline Composites. (n.d.). Retrieved March 7, 2015, from http://www.fiberline.com/structures/structural-profiles-and-brackets/why-use-composite/why-use-composite

Guide to Chinese Ceramics – The Art of Asia. (n.d.). Retrieved February 18, 2015, from http://archive.artsmia.org/art-of-asia/ceramics/early-chinese-ceramics.cfm

Johnson, R. (n.d.-a). UW Departments Web Server. Retrieved February 12, 2015, from http://depts.washington.edu/

Johnson, T. (n.d.-b). Applications of Carbon Fiber. Retrieved March 11, 2015, from http://composite.about.com/od/aboutcarbon/a/Applications-Of-Carbon-Fiber.htm

Laurencin, Cato T. "Bone Graft Substitute Materials." *eMedicine*. Eds. Jason H. Calhoun, et al. 4 Dec. 2013. Medscape. 13 May 2014 <http://emedicine.medscape.com/article/1230616-overview>.

MenuDramatica®The Next Chapter in Story Development. (n.d.). Retrieved February 12, 2015, from http://dramatica.com/analysis/the-graduate ("Menudramatica®the next chapter,")

N.A., (2012). Classification of Materials: In *Everything is Material*. Cleveland Ohio: ASM Materials Education Foundation. (Classification of Materials:, 2012)

Nitinol / Flexinol® Actuator Wire. (n.d.). Retrieved February 12, 2015, from http://www.imagesco.com/articles/nitinol/04.html

Optical fiber. (n.d.). Retrieved March 1, 2015, from http://en.wikipedia.org/wiki/Optical_fiber

Polymer regenerates all by itself. (n.d.). Retrieved February 12, 2015, from http://www.rsc.org/chemistryworld/2013/09/polymer-regenerates-elastomer-heals-independently ("Polymer regenerates all,")

R. A. Wesolowski et al., *The World of Materials*,
https://doi.org/10.1007/978-3-030-17847-5

Rabin, Steven, et al. "Immune Response to Implants." *eMedicine*. Eds. Jason H. Calhoun, et al. 4 Sep. 2013. Medscape. 13 May 2014 <http://emedicine.medscape.com/article/1230696-overview>.

Singh, Jasvinder A. "Epidemiology of Knee and Hip Arthroplasty: A Systematic Review." *Open Orthopedic Journal* 51 (2011): 80–85.

The Secret Life of Starlite. (2013, November 25). Retrieved February 12, 2015, from http://www.isciencemag.co.uk/features/the-secret-life-of-starlite/ ("The secret life," 2013)

Thomas Edison Lightbulb. (n.d.). Retrieved February 26, 2015, from http://www.edisonmuckers.org/thomas-edison-lightbulb/

Violatti, C. (2014, July 18). Stone Age. Retrieved March 19, 2015, from http://www.ancient.eu/Stone_Age/

What is composite material? definition and meaning. (n.d.). Retrieved March 7, 2015, from http://www.businessdictionary.com/definition/composite-material.html

Wilson, P. (2013, September 13). Polymer regenerates all by itself. Retrieved November 14, 2014, from http://www.rsc.org/chemistryworld/2013/09/polymer-regenerates-elastomer-heals-independently

Index

R. A. Wesolowski et al., *The World of Materials*,
https://doi.org/10.1007/978-3-030-17847-5